Marco Pilz

A comparison of proxies for seismic site conditions

Marco Pilz

A comparison of proxies for seismic site conditions

The example of Santiago de Chile

Südwestdeutscher Verlag für Hochschulschriften

Impressum/Imprint (nur für Deutschland/only for Germany)
Bibliografische Information der Deutschen Nationalbibliothek: Die Deutsche Nationalbibliothek verzeichnet diese Publikation in der Deutschen Nationalbibliografie; detaillierte bibliografische Daten sind im Internet über http://dnb.d-nb.de abrufbar.
Alle in diesem Buch genannten Marken und Produktnamen unterliegen warenzeichen-, marken- oder patentrechtlichem Schutz bzw. sind Warenzeichen oder eingetragene Warenzeichen der jeweiligen Inhaber. Die Wiedergabe von Marken, Produktnamen, Gebrauchsnamen, Handelsnamen, Warenbezeichnungen u.s.w. in diesem Werk berechtigt auch ohne besondere Kennzeichnung nicht zu der Annahme, dass solche Namen im Sinne der Warenzeichen- und Markenschutzgesetzgebung als frei zu betrachten wären und daher von jedermann benutzt werden dürften.

Coverbild: www.ingimage.com

Verlag: Südwestdeutscher Verlag für Hochschulschriften GmbH & Co. KG
Dudweiler Landstr. 99, 66123 Saarbrücken, Deutschland
Telefon +49 681 37 20 271-1, Telefax +49 681 37 20 271-0
Email: info@svh-verlag.de

Approved by: Potsdam, Universität Potsdam, Dissertation, 2010

Herstellung in Deutschland:
Schaltungsdienst Lange o.H.G., Berlin
Books on Demand GmbH, Norderstedt
Reha GmbH, Saarbrücken
Amazon Distribution GmbH, Leipzig
ISBN: 978-3-8381-2913-6

Imprint (only for USA, GB)
Bibliographic information published by the Deutsche Nationalbibliothek: The Deutsche Nationalbibliothek lists this publication in the Deutsche Nationalbibliografie; detailed bibliographic data are available in the Internet at http://dnb.d-nb.de.
Any brand names and product names mentioned in this book are subject to trademark, brand or patent protection and are trademarks or registered trademarks of their respective holders. The use of brand names, product names, common names, trade names, product descriptions etc. even without a particular marking in this works is in no way to be construed to mean that such names may be regarded as unrestricted in respect of trademark and brand protection legislation and could thus be used by anyone.

Cover image: www.ingimage.com

Publisher: Südwestdeutscher Verlag für Hochschulschriften GmbH & Co. KG
Dudweiler Landstr. 99, 66123 Saarbrücken, Germany
Phone +49 681 37 20 271-1, Fax +49 681 37 20 271-0
Email: info@svh-verlag.de

Printed in the U.S.A.
Printed in the U.K. by (see last page)
ISBN: 978-3-8381-2913-6

Copyright © 2011 by the author and Südwestdeutscher Verlag für Hochschulschriften GmbH & Co. KG and licensors
All rights reserved. Saarbrücken 2011

Content

Abstract		3
Zusammenfassung		5
1. Introduction		7
2. Tectonic framework		11
2.1.	Geological setting	12
2.2.	The San Ramón Fault and the associated hazard for Santiago de Chile	16
3. Data acquisition		19
3.1.	Temporary seismic networks	20
3.2.	Noise measurements	24
4. Comparison of site response techniques		25
4.1.	Introduction	26
4.2.	Seismic event recordings: Data acquisition and analysis	28
	4.2.1. Seismogram properties	28
	4.2.2. Time domain analysis and earthquake duration	30
	4.2.3. Denoising of seismograms using the S transform	31
	4.2.3.1. Introductory remarks	31
	4.2.3.2. Application to real data	33
	4.2.4. Frequency domain analysis	39
4.3.	H/V ratio of ambient noise: Data acquisition and analysis	43
4.4.	Results and discussion	47
	4.4.1. Seismic event and microtremor recordings: comparison of different techniques	48
	4.4.2. Summary	51
4.5.	Single station NHV measurements	53
4.6.	Fundamental resonance frequency map of the investigated area	56

	4.7.	Correlation between fundamental frequency and damage distribution of the 1985 Valparaiso and 2010 Maule events	57
5.	**3D shear wave velocity model**		**61**
	5.1.	Introduction	62
	5.2.	Inversion of H/V ratios for deriving S-wave velocity profiles	63
	5.3.	Interpolation of the Santiago basin S-wave velocity model	67
	5.4.	Characteristics and interpretation of the 3D S-wave velocity model	68
	5.5.	Correlation between slope of topography and v_s^{30}	72
	5.6.	Correlation between S-wave velocity and macroseismic intensity of the 1985 Valparaiso event	75
6.	**Simulation of the Santiago basin response by numerical modeling of seismic wave propagation**		**81**
	6.1.	Introduction	82
	6.2.	The spectral element numerical code GeoELSE	84
	6.3.	Test for accuracy and stability – the 1 April 2010 aftershock	86
		6.3.1. Implementation	86
		6.3.1.1. Mesh geometry	86
		6.3.1.2. Santiago basin model	87
		6.3.1.3. Treatment of the kinematic source	88
		6.3.2. Comparison of numerical predictions	89
		6.3.3. Effect of basin depth and surface topography	93
	6.4.	Simulating near-fault earthquake ground motion	96
		6.4.1. Implementation	96
		6.4.2. Influence of hypocenter location	97
		6.4.3. Discussion	101
7.	**Conclusions**		**103**
Appendix			107
References			109
Acknowledgements			126

Abstract

Situated in an active tectonic region, Santiago de Chile, the country´s capital with more than six million inhabitants, faces tremendous earthquake hazard. Macroseismic data for the 1985 Valparaiso and the 2010 Maule events show large variations in the distribution of damage to buildings within short distances indicating strong influence of local sediments and the shape of the sediment-bedrock interface on ground motion.

Therefore, a temporary seismic network was installed in the urban area for recording earthquake activity, and a study was carried out aiming to estimate site amplification derived from earthquake data and ambient noise. The analysis of earthquake data shows significant dependence on the local geological structure with regards to amplitude and duration. Moreover, the analysis of noise spectral ratios shows that they can provide a lower bound in amplitude for site amplification and, since no variability in terms of time and amplitude is observed, that it is possible to map the fundamental resonance frequency of the soil for a 26 km x 12 km area in the northern part of the Santiago de Chile basin.

By inverting the noise spectral rations, local shear wave velocity profiles could be derived under the constraint of the thickness of the sedimentary cover which had previously been determined by gravimetric measurements. The resulting 3D model was derived by interpolation between the single shear wave velocity profiles and shows locally good agreement with the few existing velocity profile data, but allows the entire area, as well as deeper parts of the basin, to be represented in greater detail. The wealth of available data allowed further to check if any correlation between the shear wave velocity in the uppermost 30 m (v_s^{30}) and the slope of topography, a new technique recently proposed by Wald and Allen (2007), exists on a local scale. While one lithology might provide a greater scatter in the velocity values for the investigated area, almost no correlation between topographic gradient and calculated v_s^{30} exists, whereas a better link is found between v_s^{30} and the local geology. When comparing the v_s^{30} distribution with the MSK intensities for the 1985 Valparaiso event it becomes clear that high intensities are found where the expected v_s^{30} values are low and over a thick sedimentary cover. Although this evidence cannot be generalized for all possible earthquakes, it indicates the influence of site effects modifying the ground motion when earthquakes occur well outside of the Santiago basin.

Using the attained knowledge on the basin characteristics, simulations of strong ground motion within the Santiago Metropolitan area were carried out by means of the spectral element technique. The simulation of a regional event, which has also been recorded by a dense network installed in the city of Santiago for recording aftershock activity following the 27 February 2010 Maule earthquake, shows that the model is capable to realistically calculate ground motion in terms of amplitude, duration, and frequency and, moreover, that the surface topography and the shape of the sediment bedrock interface strongly modify ground motion in the Santiago basin. An examination on the dependency of ground motion on the hypocenter location for a hypothetical event occurring along the active San Ramón fault, which is crossing the eastern outskirts of the city, shows that the unfavorable interaction between fault rupture, radiation mechanism, and complex geological conditions in the near-field may give rise to large values of peak ground velocity and therefore considerably increase the level of seismic risk for Santiago de Chile.

Zusammenfassung

Aufgrund ihrer Lage in einem tektonisch aktiven Gebiet ist Santiago de Chile, die Hauptstadt des Landes mit mehr als sechs Millionen Einwohnern, einer großen Erdbebengefährdung ausgesetzt. Darüberhinaus zeigen makroseismische Daten für das 1985 Valparaiso- und das 2010 Maule-Erdbeben eine räumlich unterschiedliche Verteilung der an den Gebäuden festgestellten Schäden; dies weist auf einen starken Einfluss der unterliegenden Sedimentschichten und der Gestalt der Grenzfläche zwischen den Sedimenten und dem Festgestein auf die Bodenbewegung hin.

Zu diesem Zweck wurde in der Stadt ein seismisches Netzwerk für die Aufzeichnung der Bodenbewegung installiert, um die auftretende Untergrundverstärkung mittels Erdbebendaten und seismischem Rauschen abzuschätzen. Dabei zeigt sich für die Erdbebendaten eine deutliche Abhängigkeit von der Struktur des Untergrunds hinsichtlich der Amplitude der Erschütterung und ihrer Dauer. Die Untersuchung der aus seismischem Rauschen gewonnenen horizontal-zu-vertikal-(H/V) Spektralverhältnisse zeigt, dass diese Ergebnisse nur einen unteren Grenzwert für die Bodenverstärkung liefern können. Weil jedoch andererseits keine zeitliche Veränderung bei der Gestalt dieser Spektralverhältnisse festgestellt werden konnte, erlauben die Ergebnisse ferner, die Resonanzfrequenz des Untergrundes für ein 26 km x 12 km großes Gebiet im Nordteil der Stadt zu bestimmen.

Unter Zuhilfenahme von Informationen über die Dicke der Sedimentschichten, welche im vorhinein schon durch gravimetrische Messungen bestimmt worden war, konnten nach Inversion der H/V-Spektralverhältnisse lokale Scherwellengeschwindigkeitsprofile und nach Interpolation zwischen den einzelnen Profilen ein dreidimensionales Modell berechnet werden. Darüberhinaus wurde mit den verfügbaren Daten untersucht, ob auf lokaler Ebene ein Zusammenhang zwischen der mittleren Scherwellengeschwindigkeit in den obersten 30 m (v_s^{30}) und dem Gefälle existiert, ein Verfahren, welches kürzlich von Wald und Allen (2007) vorgestellt wurde. Da für jede lithologische Einheit eine starke Streuung für die seismischen Geschwindigkeiten gefunden wurde, konnte kein Zusammenhang zwischen dem Gefälle und v_s^{30} hergestellt werden; demgegenüber besteht zumindest ein tendenzieller Zusammenhang zwischen v_s^{30} und der unterliegenden Geologie. Ein Vergleich der Verteilung von v_s^{30} mit den MKS-

Intensitäten für das 1985 Valparaiso-Erdbeben in Santiago zeigt, dass hohe Intensitätswerte vor allem in Bereichen geringer v_s^{30}-Werte und dicker Sedimentschichten auftraten.

Weiterhin ermöglichte die Kenntnis über das Sedimentbeckens Simulationen der Bodenbewegung mittels eines spektralen-Elemente-Verfahrens. Die Simulation eines regionalen Erdbebens, welches auch von einem dichten seismischen Netzwerk aufgezeichnet wurde, das im Stadtgebiet von Santiago infolge des Maule-Erdbebens am 27. Februar 2010 installiert wurde, zeigt, dass das Modell des Sedimentbeckens realistische Berechnungen hinsichtlich Amplitude, Dauer und Frequenz erlaubt und die ausgeprägte Topographie in Verbindung mit der Form der Grenzfläche zwischen den Sedimenten und dem Festgestein starken Einfluss auf die Bodenbewegung haben. Weitere Untersuchungen zur Abhängigkeit der Bodenerschütterung von der Position des Hypozentrums für ein hypothetisches Erdbeben an der San Ramón-Verwerfung, welche die östlichen Vororte der Stadt kreuzt, zeigen, dass die ungünstige Wechselwirkung zwischen dem Verlauf des Bruchs, der Abstrahlung der Energie und der komplexen geologischen Gegebenheiten hohe Werte bei der maximalen Bodengeschwindigkeit erzeugen kann. Dies führt zu einer signifikanten Zunahme des seismischen Risikos für Santiago de Chile.

Chapter 1

Introduction

Extended mountain valleys with wide plains of fluvial deposits or lakeshores and estuaries with water-saturated sediments are particularly prone to seismic site amplification and non-linear effects. In former times, such seismically unfavourable sites were attractive for spacious settlements and industries, and many cities worldwide have grown extensively over such plains and are still expanding. Given this spread of urban populations into areas of unfavorable soils all over the world, future earthquakes might cause extensive human and economic damage, since geometrical and mechanical features of alluvial deposits have great influence on seismic wave propagation and amplification as seen from many recent events (Ansal 2004). The Michoacan, Mexico, event (1985, Mw=8.1) and the Hyogo-ken Nanbu, Japan, earthquake (1995, Mw=6.9) can serve as notable examples of the consequences of such site effects. Moreover, many of these urban settlements worldwide are located in tectonically active regions which are particularly threatened by large earthquakes.

Both issues also apply to the city of Santiago de Chile, the country's rapidly growing capital with more than six million inhabitants. On the one hand, Chile is one of the most seismically active areas in the world. Subsequently, this sees also the city of Santiago confronted with a high level of seismic hazard. Due to the subduction of the oceanic Nacza Plate beneath the continental South American lithosphere, leading to a convergence rate of about 6 to 7 cm per year around central Chile (Khazaradze and Klotz 2003), a number of destructive earthquakes has occurred. Within the last 100 years, 16 earthquakes have caused significant human and economic losses in Chile. Return periods for magnitude 8 events are of the order 80-130 years for any given region in Chile, but they decrease to approximately 12 years when the country as a whole is considered (Barrientos et al. 2004).

On the other hand, the city of Santiago de Chile is located in a narrow basin between the Andes and the coastal mountains, filled with soft sediments that may strongly modify ground motion. The geometry of the subsoil structure, the soil types and the variation of their properties with depth, lateral discontinuities and the surface topography are at the origin of large lateral variances of ground motion. Also for the Santiago basin pronounced variations of soil conditions and topography can be found over relatively short distances which led to intensities varying between VII and IX (MSK scale) within the city during the 1985 Valpariso event (Çelebi 1987, Bravo 1992). In contrast to this there is the lack of recording of ground motion behaviour for sites in the Santiago basin.

For seismic hazard assessment and mitigating earthquake disaster on local scale it is therefore keenly required to develop methods which are able to characterize site effects and help to understand soil behavior. Thus, assessing the amplification / de-amplification and lengthening of ground motion due to surficial geology and topography is of primary importance for earthquake microzonation studies and has led to the development of systematic approaches for mapping seismic site conditions (e.g. Park and Elrick 1998, Wills et al. 2000, Holzer et al. 2005), as well as quantifying both amplitude- and frequency-dependent site amplifications (e.g. Borcherdt 1994). In recent years, due to increasing computational power, also 3D numerical simulations have been successfully applied to study the complex nature of strong ground motion (e.g. Olsen et al. 1995, Wald and Graves 1998, Komatitsch et al. 2004, Lee et al. 2008, Stupazzini et al. 2009).

However, because of the growing importance of reliable seismic hazard assessment in large urban areas much effort has been spent in the development of tools for site effect estimation that allow large areas to be covered at limited costs. A now standardized approach for mapping seismic site conditions is measuring and mapping the average shear wave velocity in the uppermost 30 m (v_s^{30}). The Uniform Building Code (International Conference of Building Officials 1997) uses v_s^{30} to group sites into several broad classes, whereas each category is assigned by characteristic factors that can modify the response spectrum, i.e. the response of the significant eigenmodes of the structure, in a different way. The U.S. Building codes (Building Seismic Safety Council 2004) as well as the Eurocode 8 (CEN 2003) now also rely on v_s^{30} for seismic site characterization. Recently, Wald and Allen (2007) presented a method to derive site-condition maps by correlating v_s^{30} with the topographic gradient. The basic conclusion of this technique is that the topographic slope from both high and low resolution maps might be used as a reliable indicator of v_s^{30} in the absence of geologically- and geotechnically-based site-condition information, since more competent materials, which are characterized by higher seismic velocities, would be more likely to maintain higher gradients, whereas soft sediments are deposited predominantly in areas with a low gradient.

To check the potential of the previously introduced technique on a local level, the city of Santiago de Chile would be an appropriate test area. In fact, only few investigations have been conducted in Santiago de Chile to evaluate site effects; these studies predominantly deal with microtremor measurements (Toshinawa et al. 1996, Bonnefoy-Claudet et al. 2009) and, therefore, provide only a limited amount of information. On the other hand, Cruz et al. (1993) accounted for data from several earthquakes in their site effect study, but most of the stations were located on the outskirts of the city and only a few, and very local, events were considered.

The presented work focuses both on experimental as well as on numerical aspects of seismic hazard assessment, whereas the coupling to seismic risk is discussed briefly only for a few cases. In particular, it will be tested, if the slope of topography may serve as a reliable proxy for the evaluation of seismic velocities and therefore be appropriate for site response estimation. For this purpose, chapter 2 describes the geological settings in the Santiago de Chile basin and the seismic hazard in central Chile. Chapter 3 gives a short overview of the field experiments carried out. Chapter 4 illustrates a detailed site effect analysis using data both from earthquakes recorded by a temporary seismic network installed in the northern part of the city and from microtremor measurements. As the modification of strong ground shaking is indeed related to the S-wave velocity structure below the site, it is further possible to check if any correlation between the parameters of topographic gradient, v_s^{30}, and surface geology may exist. Results of this investigation are summarized in chapter 5. Moreover, the wealth of the available data further allowed performing 3D numerical simulations of ground motion in the Santiago basin. In chapter 6 waveform recordings of aftershocks following the 27 February 2010 Maule earthquake are compared with numerical simulations. Furthermore, the variability of ground motion in the Santiago de Chile basin is assessed by analysing the results of different earthquake scenarios.

Chapter 2

Tectonic framework

2.1. Geological setting

At the forearc of central Chile (33-34°S) three morphostructural units are recognizable parallel to each other and oriented north-south (Figure 2.1). This configuration was generated during a period of maximal compression during the upper Oligocene-Pliocene period (Thiele 1980).

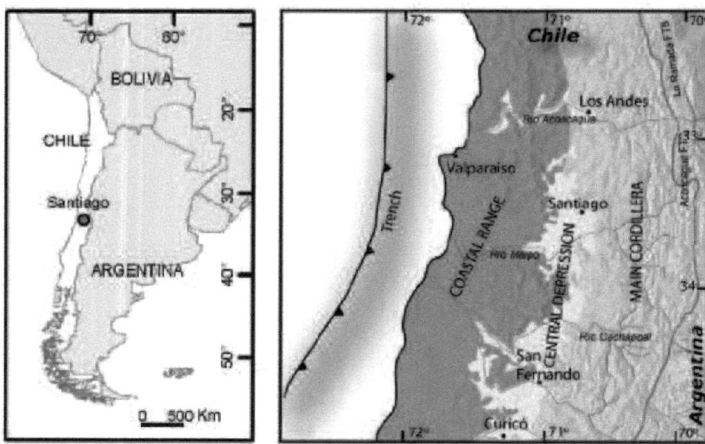

Figure 2.1: Regional morphostructural map of central Chile, showing the tectonic setting and regional geomorphological units.

Located within the central depression, the Santiago basin is around 80 km long, 40 km wide and mainly aligned in north-south direction (Figure 2.2), limited by a watershed to the north by the El Manzo cordillera and to the south by the hills of Angostura. The basin originated from the depression caused by tectonic movements in the Tertiary of an area between two major faults running parallel to the two mountain chains.

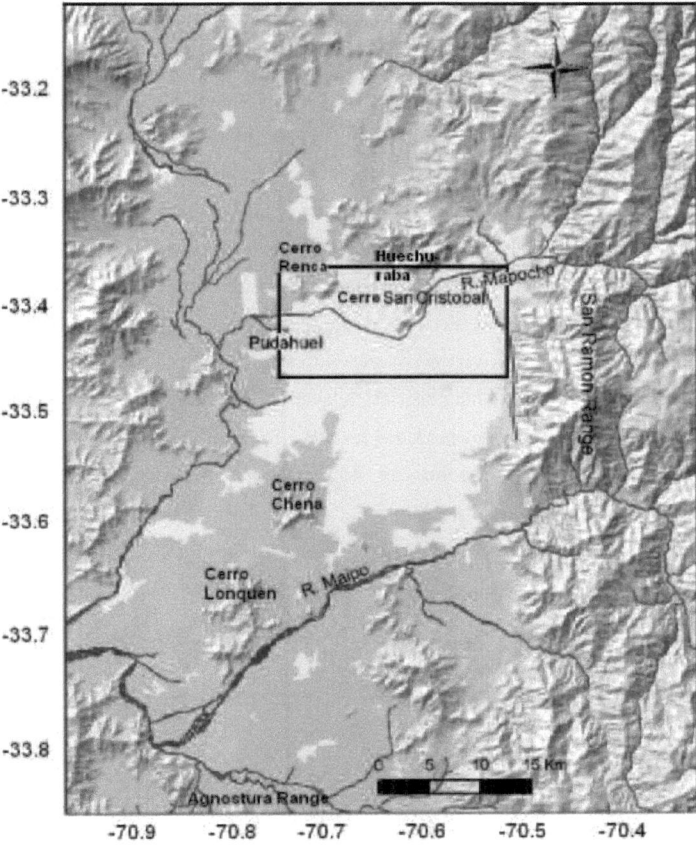

Figure 2.2: Basin of Santiago de Chile. Areas of high housing density are shown in salmon-red. Locations mentioned in the text are indicated. The red line represents the San Ramón Fault (see chapter 2.2). The area of investigation is marked by a black rectangle.

The surface geology of the Santiago basin consists of several different materials, as shown in Figure 2.3 (Valenzuela 1978). The city center and most of the north-eastern parts of the city are located on fluvial gravel deposits (Santiago gravel and Mapocho gravel in Figure 2.3). This kind of soil is composed of boulders usually less than 20 cm in diameter in a matrix that varies from silty gravel to silty sand with sandy and clayey lenses. The soil has a high density and a low deformability. It originates in the deposits of the river Mapocho (and to a lesser part of the river Maipo), grading to coarser deposits closer to the apex of the rivers' alluvial fans, i.e. to the north-east. Alluvial silt and high plasticity clay soils deposits (fine-grained soil in Figure 2.3) make up most of the north-western part of the investigated area, with a transition zone existing between these two areas. The south-eastern part of the basin is mainly dominated by the alluvial fans of the ravines that drain the San Ramón range (mudflow deposits in Figure 2.3). These soils are composed by rock blocks in a silty and clayey matrix with varying amounts of sand, deposited by alluvial and hillslope processes, and present heterogeneous geotechnical properties: closer to the mountain foothills, the materials are coarser and more heterogeneous, while in the distal zones they tend to be more stratified and homogeneous in texture and granulometry. The riverside of the Mapocho river in the central part of the city is covered with artificial fill. In the south-western part of the investigated area (Pudahuel district), a layer of ash (pumice) is known to be at the top of the sedimentary column, probably resulting from a major eruption of the Maipo volcano, located 120 km to the south-east, 450 000 years ago (Stern et al. 1984, Baize et al. 2006). One must, however, keep in mind that the basin geometry, characterized by sharp lateral and vertical variations of various geological units, is rather complex, resulting in a high level of uncertainty in our knowledge and has not been mapped in more detail so far.

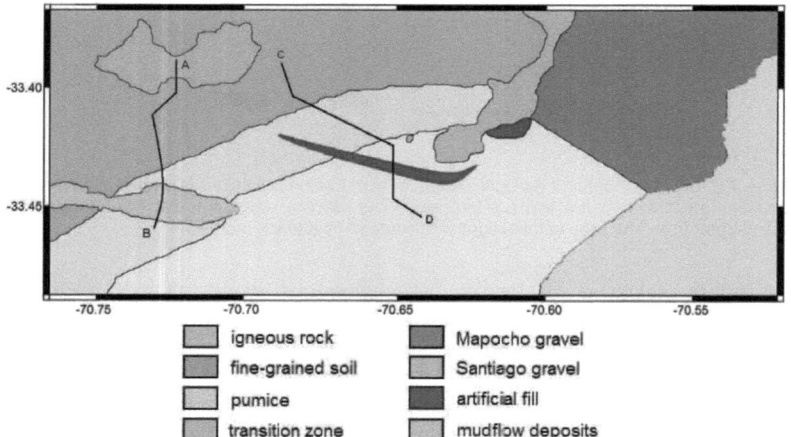

Figure 2.3: Simplified map of surface geology of the investigated area (after Valenzuela 1978). Thick lines mark the tracks of the cross sections shown in Figure 5.2.

The basement of the Santiago basin is believed to result from volcanic activity aged between the higher Oligocene to the lower Miocene (igneous bedrock, Abanico formation). The bottom of the basin, only known indirectly from gravimetric measurements (Araneda et al. 2000), corresponds to an uneven surface that conceals some hill islands that locally outcrop within the basin (Figure 2.4). The alignment of some of these hills, such as that for San Cristóbal, Chena, and Lonquén, suggest that there might have been a structural control.

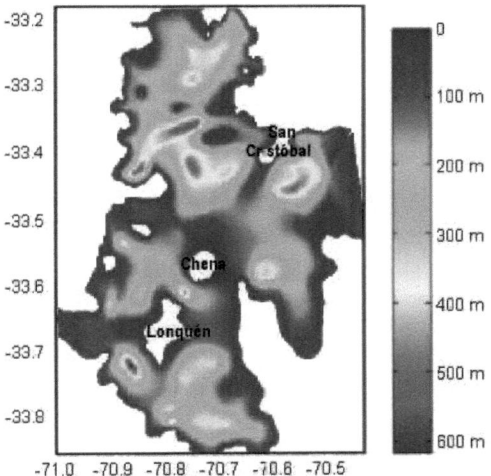

Figure 2.4: Thickness of the sedimentary cover of the investigated area as determined by interpolation of gravimetric data (Araneda et al. 2000). White areas in the middle of the basin can be identified as outcropping hills.

The thickness of the sedimentary cover varies over short scales and can exceed more than 600 m but has been assessed directly (by drilling) only to approximately 120 m depth. However, the basin geometry is rather complex, therefore having a strong influence on the frequency band affected by amplification.

2.2. The San Ramón Fault and the associated hazard for Santiago de Chile

The West Andean Front near Santiago appears as a major tectonic contact between the Main Cordillera, which corresponds to the pervasively shortened eastern side of the Andean Basin, a deep, very long (several 10^3 km long) and relatively narrow (of the order of tens of kilometers wide) basin with significant westward dip nearly parallel to the Andes between the equator and latitude 48° S (Mpodozis and Ramos 1989, Vicente 2005), and the marginal block, which constitutes the relatively shallow dipping western side of the Andean Basin (Thiele 1980).

Herein, the San Ramón Fault represents a multikilometric frontal thrust at the western front of the Main Cordillera; it is interpreted as a growing west vergent fault propagation fold system (Rauld et al. 2008, Armijo et al. 2010). Its basal detachment is close to the base of the Andean Basin at ~12 km of stratigraphical depth. The structure associated with the San Ramón Fault is well constrained by the mapped surface geology and can be restored to deduce the amounts of shortening and uplift. Total shortening across the frontal system is around 10 km. It has occurred since ~25 Ma, according to precise dates in the Abanico formation (Armijo et al. 2010). The shortening, uplift and erosion rates across the Main Cordillera are distributed inhomogeneously; no extensive erosion surface has developed.

The San Ramón Fault reaches the surface with steep eastward dip. The minimum average slip rate would be of the order of 0.2 mm per year and the slip rate on the basal detachment of the frontal system of the order of 0.4 mm per year. However, the growth process of the thrust front suggests that the most of the slip on the basal detachment has localized since 16 Ma in the San Ramón Fault, making of it the frontal ramp of the Main Cordillera. Thus, the actual long term average slip rate on the San Ramón Fault would be of around 0.3 mm per year (Armijo et al. 2010).

The best surface expression of the San Ramón Fault is found along an around 15 km long segment with a sharp fault trace in the 25 km long part separating the rivers Mapocho and Maipo along the San Ramón mountain front (see Figure 2.2). The occurrence of ash lenses correlated with the Pudahuel pumacit in the piedmont yields a strictly minimum throw rate of 0.13 mm per year (≥ 60 m in 450,000 years since the eruption of the Maipo volcano). Throw of about 4 m was measured across a well preserved scarp that appears to be the last testimony to late scarp increments left for study along the San Ramón Fault. This feature is to be accounted for by a single event or by several events with thrust slip of the order of ~1 m or less. When using a conservative estimate of the average slip of 1 to 4 m, following Armijo et al. (2010), who excluded seismic creep in their study, a shear modulus of 30 GPa with a rupture of the entire length of the San Ramón mountain front facing the Santiago valley (~30 km) and further including well constrained seismogenic zones down to 15 km depth under the Principal Cordillera, this will yield seismic moments of 1.4 to $5.4 \cdot 10^{19}$ Nm, corresponding to events of magnitude Mw=6.7 to Mw=7.1. This range of magnitudes is significantly higher than that of the sequence of three consecutive shocks, all together known as the 1958 Las Melosas earthquake, which correspond to the largest events recorded instrumentally in the upper plate near Santiago (see Sepúlveda et al. 2008, Alvarado et al. 2009 and references therein). The 1958 sequence of the three mainshocks occurred within an interval of 6 min with hypocentral depth of 10 km in the center of the Main Cordillera 60 km south-east of Santiago, with intensity values reaching IX to X (MKS scale) in the epicentral area. The larger first shock has been assigned a revised magnitude Mw=6.3 (Alvarado et al. 2009).

However, earthquakes with large magnitudes would not be too frequent (of the order of 2,500 to 10,000 years), because the loading rate of the San Ramón Fault seems to be low. Thus, the probability of having recorded such an event is low, given the short period of settlement in the Santiago basin (the city was founded in 1541). Nevertheless, events that large could not be disregarded for seismic hazard assessment in the Santiago region and are therefore considered when discussing different earthquake scenarios in chapter 6.

Chapter 3

Data acquisition

3.1. Temporary seismic networks

In 2008, a network of eight seismological EarthData Logger PR6-24 instruments equipped with a Mark L-4C-3D sensor was deployed in the northern part of Santiago de Chile, covering the different geological units (Table 3.1, see also Figure 3.1). We used 1 Hz short period sensors which had been found to be suitable for investigation also when the fundamental frequency of soils is expected to occur down to 0.1 to 0.2 Hz (Strollo et al. 2008a, Strollo et al. 2008b). Installation of the network started on 20 March 2008, and the stations were removed after 26 May 2008, with exception of station S8 which had been removed some days before. A brief description of the soil conditions at the sites of stations installed for a short term based on borehole data and previous investigations compiled by Valenzuela (1978), Araneda et al. (2000) and Rebolledo et al. (2006) is given below. Note that there is a degree of uncertainty in the depth of the bedrock due to interpolation and the inversion procedure of the gravimetry data between the measurement sites which were spaced with an interstation distance of around 500 m.

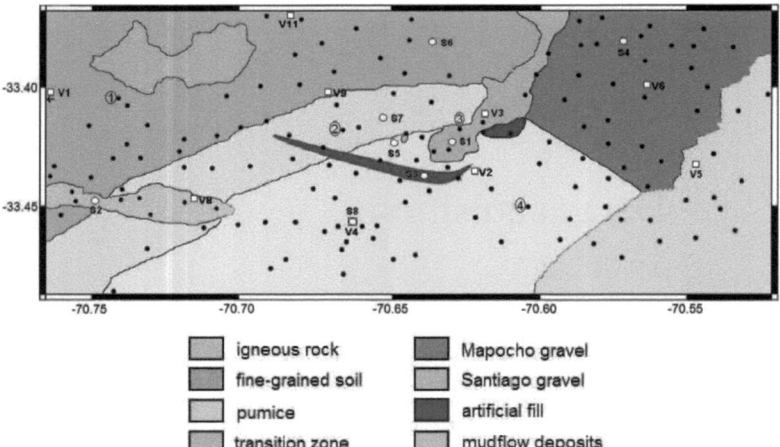

Figure 3.1: Geological map of the area of investigation with large white spots indicating the locations of the seismic station installed in 2008 (Table 3.1). White quads represent locations of the seismic stations installed in 2010 (Table 3.2) whose recordings are used for further analyses. Small black spots show sites where measurements of seismic noise have been carried out. Encircled numbers indicate locations of S-wave velocity profiles shown in Figure 5.1.

Table 3.1: Locations of seismic network (March to May 2008).

station number	latitude	longitude	location	installation	geological unit
S1	-33.424	-70.632	hut on San Cristóbal hill	at ground level	igneous rock
S2	-33.444	-70.745	aeronautic institute	basement	pumice
S3	-33.435	-70.644	museum Bellas Artes	cellar	artificial fill
S4	-33.384	-70.573	library of Bradford college	cellar	Mapocho gravel
S5	-33.425	-70.645	cemetery General	basement	Santiago gravel
S6	-33.388	-70.634	cemetery Parque del recuerdo	at ground level	fine-grained soil
S7	-33.418	-70.655	faculty of medicine	cellar	transition zone
S8	-33.457	-70.663	faculty of physics	cellar	Santiago gravel

Station S1: Igneous rock (outcropping San Cristóbal hill) composed of andesite and granodiorite. The San Cristóbal hill rises almost 300 m above the surrounding plain. The hut housing the instrument is approximately 30 m below the top of the hill. Since S1 is situated on rock it can serve as a reference station, although topographic effects and further perturbing influences cannot be excluded as will be discussed in the following.

Station S2: Pumacit deposits consisting of volcanic ash and pumice accompanied by fragments of rocks that are occasionally included in the overlaying gravel layer. The thickness of the pumacit layer is around 45 m, with the depth to the bedrock being around 380 m.

Station S3: Heterogeneous materials ranging from sand to gravel rubble and waste. The soil shows poor quality as a foundation, as there is no control regarding the composition and the degree of compaction. Depth to the bedrock is assumed to be around 80 m.

Station S4: Alluvial deposits of the Mapocho river consisting of very dense gravel and unsaturated soil accompanied by sandy gravel, sand, silt and clays. Depth to the bedrock is assumed to be around 90 m.

Station S5: Santiago gravel derived from alluvial from the river Mapocho and to a lesser part from the river Maipo. The soil is characterized by large gravel compactness in a sand-clay matrix. Interpolation of gravimetric data suggest that the depth of the bedrock is around 185 m, but this might be questionable since the station is located close to an outcrop of the small Cerro Blanco. The thickness of the sedimentary cover decreases rapidly when approaching this outcrop, but this cannot be seen in the gravimetric data.

Station S6: Fine-grained soil of Santiago's northwest of silt and inorganic clays of high plasticity with thin horizons of gravel and volcanic ash. Depth to the bedrock is around 136 m.

Station S7: Transition zone between Santiago gravel and fine-grained soil including thick clay material layers. Large horizontal and vertical stratigraphic variations can be found within this zone. Depth of the bedrock is around 177 m.

Station S8: Santiago gravel from the river Maipo alluvium and to a lesser degree from the river Mapocho. The soil is characterized by large gravel compactness in conjunction with clay and a lesser amount of sand. Depth of the bedrock is around 95 m.

Moreover, in immediate succession of the 27 February 2010 Maule (Chile) earthquake (Mw=8.8) a network of twelve EarthData Logger instruments equipped with a Mark L-4C-3D sensors and eight accelerometers Kinematrics Altus K2 was installed in the city of Santiago for recording aftershock activity (Table 3.2, see Figure 3.1). Installation of the network started on 11 March 2010 and was completed on 18 March 2010. Removal of the stations began on 20 May 2010. Although a detailed analysis of the recorded data has not been completed yet, a small fraction of the recordings (around 10 %) will be considered for further analysis.

Table 3.2: Locations of seismic network (March to May 2010). Herein, station numbers with V represent velocimeters, station numbers with A represent accelerometers.

station number	latitude	longitude	location	installation	geological unit
V1	-33.403	-70.780	storehouse at the airport	at ground level	fine-grained soil
V2	-33.433	-70.624	courtyard of hotel Presidente	at ground level	Santiago gravel
V3	-33.412	-70.621	police dog obedience school	free field	igneous bedrock
V4	-33.457	-70.662	geophysical institute	cellar	Santiago gravel
V5	-33.429	-70.548	condominium Las Condes	cellar	mudflow deposits
V6	-33.399	-70.568	German school Las Condes	at ground level	Mapocho gravel

V7	-33.442	-70.683	faculty of medicine Quinta Normal	cellar	Santiago gravel
V8	-33.443	-70.718	municipality Lo Prado	at ground level	pumice
V9	-33.396	-70.672	municipality Conchalí	at ground level	fine-grained soil
V10	-33.377	-70.509	headmaster German school	free field	Mapocho gravel
V11	-33.360	-70.687	municipality Huechuraba	at ground level	fine-grained soil
V12	-33.438	-70.750	hospital Pudahuel	at ground level	fine-grained soil
A1	-33.399	-70.568	German school Las Condes	at ground level	Mapocho gravel
A2	-33.499	-70.611	faculty of physics	cellar	Santiago gravel
A3	-33.435	-70.635	faculty of law	cellar	artificial fill
A4	-33.419	-70.656	faculty of medicine Independencia	at ground level	transition zone
A5	-33.375	-70.635	municipality Huechuraba	at ground level	fine-grained soil
A6	-33.404	-70.704	municipality Renca	at ground level	fine-grained soil
A7	-33.445	-70.704	municipality Pudahuel	cellar	pumacit
A8	-33.423	-70.743	police department Cerro Navia	at ground level	fine-grained soil

During both campaigns the short-term stations were installed inside buildings and were placed in their room´s corner and, whenever possible, in cellars. (Two stations of the 2010 network were installed at shielded sites in free field.) All stations were synchronized using GPS reference time, however, station S8 lost its GPS signal some time after installation. While this might affect the localization of earthquakes, there is obviously no influence on the spectral shapes. The signal was recorded at each site with a sampling rate of 100 samples per second. Due to the high level of noise within the urban area, which increases the chances of either false triggering as well as the risk of not triggering on the weak earthquakes, all velocimetric stations were set to continuous recording. Therefore, in addition to the recorded earthquakes, a huge amount of ambient noise data was recorded. The accelerometric stations of the 2010 network were operating in a triggering mode.

3.2. Noise measurements

From 19 May until 13 June 2008 and from 19 to 23 May 2010, noise measurements were carried out in the northern part of Santiago de Chile using the same setup described above. At each site the signal was recorded for at least 25 minutes, leading to 146 measurements of ambient noise being carried out (measurement sites are shown in Figure 3.1). For almost all the measurements in 2008 the sensor was placed directly on the ground (i. e. mown grass, not overgrown soil), with only few measurements carried out on asphalt. For all the measurements the sensor was protected against wind. On some days rain had been slight to moderate, but it has been shown (Chatelain et al. 2008) that rain has no noticeable influence. The measurements in 2010 have been carried out within selected representative buildings, both damaged and undamaged, for the determination of characteristic building parameters.

Chapter 4

Comparison of site response techniques

4.1. Introduction

To mitigate the risk associated with the recurrence of earthquakes and site effects, a number of procedures suitable for mapping the mechanical properties of the ground near the surface and to assess level of ground shaking are currently used. Nowadays, experimental determination of the site response is generally accomplished by the spectral ratio method using a reference station (e.g. Bard and Riepl-Thomas 2000). Due to the fact that it has provided consistent results (Field and Jacob 1995, Bonilla et al. 1997, Parolai et al. 2000, Frankel et al. 2002), the standard spectral ratio technique (SSR, Borcherdt 1970) is most commonly used. One important precondition for using the SSR technique is the availability of a reference (bedrock) site with negligible site response close to the considered soil site. As this may not always be possible, other methods that can overcome this limitation have been proposed. For example, the single horizontal-to-vertical (H/V) technique requires one station recording only and uses the vertical component as a reference. The method is a combination of the receiver-function technique proposed by Langston (1979) and the method of Nakamura (1989). Langston's approach is based on the basic assumption that the vertical component is relatively uninfluenced by local geological structure. The method of Nakamura (1989), based on the approach of Nogoshi and Igarashi (1970, 1971), defined the site response as the ratio of the horizontal to vertical motion at the surface by assuming the vertical component is not amplified by the surface layers. Although Nakamura's approach was originally used for microtremors, it has also been extended to earthquake recordings, first applied to earthquake S-waves by Lermo and Chavez-Garcia (1993).

Usually, the H/V ratio from microtremors shows a clear peak in good agreement with the fundamental resonance frequency at soft soil sites under the constraint of the existence of a large impedance contrast between the sediments and bedrock (Lermo and Chavez-Garcia 1994, Field and Jacob 1995, Horike et al. 2001, Bard 2004). In contrast, the peak amplitude of the microtremor ratio often tends to underestimate the peak amplitude of earthquake SSR (e.g. Field and Jacob 1995, Bindi et al. 2000, Parolai et al. 2004a, Bard 2004, Haghshenas et al. 2008); only in few studies a close match between both amplitudes is found (e.g. Horike et al. 2001, Mucciarelli et al. 2003, Molnar and Cassidy 2006). Therefore, it is in general agreement that the NHV technique can provide the fundamental frequency and a lower-bound estimate of the amplification for a soft soil site, but fails to provide higher harmonics. However, the meaning of the amplitude of the peak are still a topic of discussion.

Several site effect studies based on the H/V method for both earthquakes (EHV) and microtremors (NHV), as well as the SSR method, representing results from various sedimentary basins, have been published over the last few years. Although their geological settings are different, deep basins filled with unconsolidated Quaternary deposits several hundred meters thick dominate these studies. Therefore, the simple 1D resonance frequency of the sediment columns is relatively low, such as in the case of the Cologne area (Parolai et al. 2004a), the Lower Rhine embayment (Ibs-von Seht and Wohlenberg 1999), the Western part of the city of Basel located in the Rhine graben (Fäh et al. 1997) and the Rhone valley in Valais in Switzerland (Frischknecht et al. 2005), Grenoble (Lebrun et al. 2001) in France, Thessaloniki (Panou et al. 2005) and the Volvi basin (Theodulidis 2006) in Greece, as well as the Molise basin (Gallipoli et al. 2004a) in Italy. Examples of shallow and thinner deposits, resulting in a higher resonance frequency, are also reported for the area of Lisbon (Teves-Costa et al. 1996) in Portugal, in the Umbria-Marche region (Mucciarelli and Monachesi 1998) in Italy, the Victoria basin (Molnar and Cassidy 2006) in British Columbia, and the Bovec basin (Gosar 2008) in Slovenia.

Since a clear understanding of the H/V ratio derived from ambient noise measurements is still lacking, the observations should be checked for each area under investigation. Therefore, a detailed site effect study will be carried out by combining data from both earthquake recordings made by a temporary seismic network described in chapter 3.1.1. and from measurements of ambient seismic noise (chapter 3.1.2.). After providing a detailed analysis of the recorded events both in time and in the frequency domain, three different methods for obtaining site responses will be compared: H/V from earthquake recordings, the standard spectral ratio technique as well as H/V from microtremors. Remarks about the benefits and caveats of the different site response techniques will also be included. Additionally, the stability of the NHV in terms of time and amplitude will be checked. As a result of the analysis of NHV measurements carried out in the northern part of Santiago de Chile, a map of the fundamental resonance frequency of the investigated area will be provided; moreover, the results are compared with observations of damaged structures in the city following the 2010 Maule earthquake.

4.2. Seismic event recordings: Data acquisition and analysis

4.2.1. Seismogram properties

While the network had been installed in 2008, 38 earthquakes were recorded with sufficient quality. Table A.1 shows all the hypocentral parameters and magnitudes according to the PDE catalogue. The list includes only earthquakes with an epicentral distance to the network of at least five times the maximum distance to the reference station, which is 11 km between stations S1 and S2, to exclude path effects. Altogether, ten local, 22 regional and six teleseismic events are listed. Almost all earthquakes had been recorded by all stations, with an exception of station S4 by which only 14 events with sufficient quality were recorded (see Table A.1).

As an example, Figure 4.1 shows the unfiltered recordings of the NS component for all stations of one local event. It can be seen that stations S2 and S6, which were located close to busy roads, were affected by higher noise amplitudes while, e.g. for stations S1, the reference station, and S5, the latter being located in a quiet part of a cemetery, the noise levels are much lower.

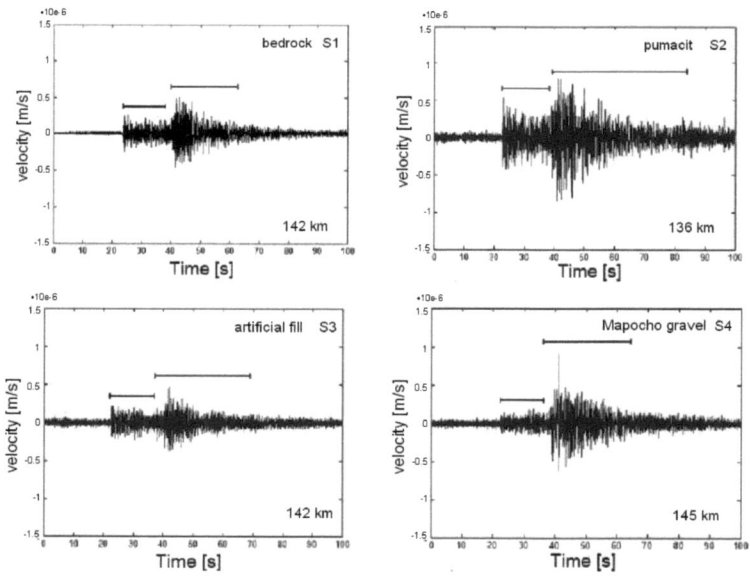

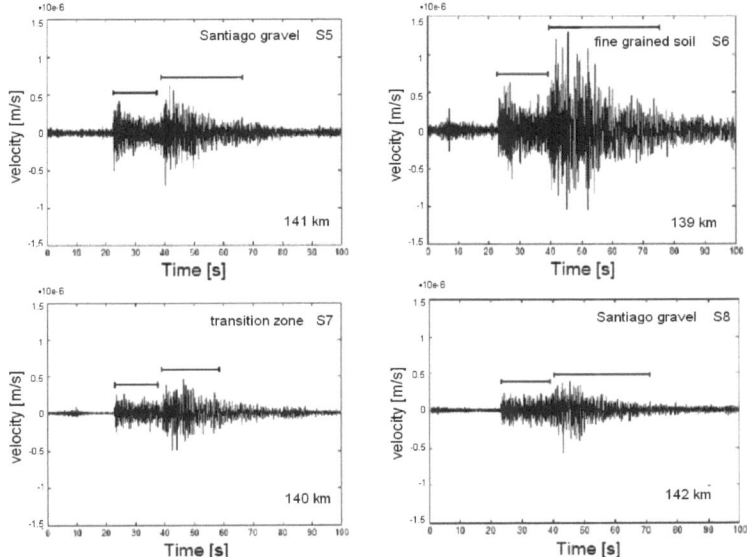

Figure 4.1: Comparison of NS recordings of a local earthquake on 12 April 2008, 21:21:57 UTC. Some stations show higher noise level due to the environment where they had been installed. The station ID (cf. Figure 3.1 and Table 3.1) and the hypocentral distances, as well as the time windows for P-wave and S-wave analysis are indicated.

4.2.2. Time domain analysis and earthquake duration

Quite different amplitudes and durations are observed at separate stations despite their proximity. While the amplitude ratio can differ by more than a factor of two compared to reference station S1 (see e.g. station S6), Figure 4.1 also shows clearly the lengthening of ground shaking for the basin sites. We calculate the duration of an event by ignoring the first and the last five per cent of the velocity square integral and considering the remaining 90 per cent as the significant contribution (Trifunac and Brady 1975). Several other definitions of earthquake durations can be found in a review by Bommer and Martínez-Pereira (1999).

The durations of all recorded events were divided by the corresponding duration recorded at the reference station S1. Results of the mean duration ratios and standard deviations are shown in Table 4.1. As can be seen, the duration of ground shaking is increased on average by a factor of nearly 2, with a maximum of 2.27 at S6; only at station S5 the duration ratio is close to 1. This might be due to the location of the station close to (approx. 350 m) the outcrop of igneous bedrock at Cerro Blanco. Only a small sedimentary cover of stiff gravel underlies S5 resulting in a lower amplification and ground motion duration. As already seen in Figure 4.1, station S6 not only shows largest amplitudes but also on average most significant lengthening of the duration.

Due to the large standard deviations that can be found in Table 4.1, we also performed an F-test. Although the nullhypothesis of equal distribution at 95 % confidence level must not be rejected for the present data set, meaning that there is a statistically significant correlation between the lengthening of duration for different stations, we are anyhow quite sure that a more comprehensive data set should reduce the large standard deviations.

Table 4.1: Duration ratios and standard deviations for stations located in the basin divided by the duration of the reference station S1.

station	duration ratio	σ
S2	1.89	0.83
S3	1.91	1.05
S4	1.65	0.44
S5	1.21	0.34
S6	2.27	1.21
S7	1.75	0.59
S8	1.59	0.70

In Figure 4.2, the duration ratios calculated on the basis of both local, regional, and teleseismic earthquakes for stations S5 and S6 relative to the reference station S1 as a function of back azimuth and magnitude are shown. In general, a large scattering of duration lengthening in the middle of the basin is observed. However, the recorded earthquakes do not cover the entire back azimuth range and therefore do not allow to assess, if there is any clear dependence between the location of the event and the observed duration. On the other hand, a slight correlation between the extension of duration and the event's magnitude rather than the epicentral distance seems to exist: local events occuring with a lower magnitude (squares in Figure 4.2) show only a slight lengthening of the duration ratio, whereas for regional and teleseismic events, a significant increase in duration ratio is observed, an expected result. This might be explained by considering the different types of waves and the resulting incident angles arising from local, regional, and teleseismic events.

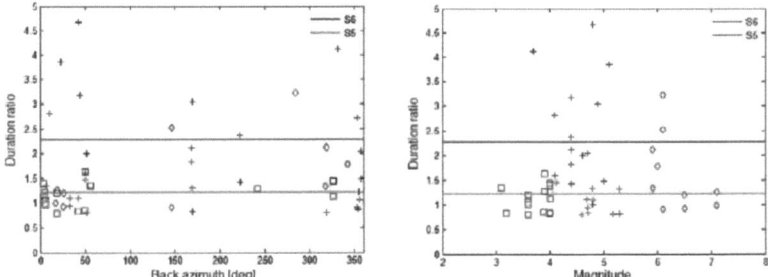

Figure 4.2: Duration ratios calculated for stations S5 and S6 with respect to the duration observed at station S1. The ratios are shown against the angle of the back azimuth (left) and magnitude of the event (right). Squares, crosses, and diamonds represent local, regional, and teleseismic events, respectively. The horizontal lines correspond to the mean duration ratios for stations S5 (red) and S6 (black).

However, since the data were acquired in urban areas some recordings are strongly affected by noise. Hence, there is a strong need of suitable procedures that allow a reliable identification of different phases without losing important information, i.e. the necessity of tools that can effectively take into account the frequency-time variation of the seismic recordings.

4.2.3. Denoising of seismograms using the S transform
4.2.3.1. Introductory remarks

Already in the 1980s, the existence of basin-induced surface waves in the coda of S-waves was demonstrated through numerical simulation for two-dimensional basin models (e.g. Bard and Bouchon 1980, Vidale and Helmberger 1988, Kawase and Aki 1989) and was also pointed out by several observational studies in the same decade. For example, Idei et al. (1985) suggested the contamination of surface waves in late S-coda based on the fact that phase velocities estimated from late S-coda observed by a tripartite array in the Kyoto basin, Japan, agree with theoretical ones for Rayleigh waves in the frequency range from 0.75 to 3 Hz and that the observed amplitudes of late S-coda are larger than theoretical ones obtained from a one dimensional model, which can reproduce observed amplitudes of an S-wave in the time domain. Phillips et al. (1993) showed that phase velocities estimated from late S-coda observed by a horizontal array in the Kanto basin, Japan, agree with theoretical ones for Love waves at around 1 Hz and that these Love waves seem to be generated at the edge of the Kanto basin.

However, direct observations of such basin-edge-induced waves are still relatively rare and often limited to large size structures (more than 50 kilometers) where the late edge-generated surface wave phase can be easily seen several tens of seconds after the S-wave train (e.g. Kagawa et al. 1992, Kawase and Sato 1992, Kinoshitaet al. 1992, Phillips et al. 1993, Frankel 1994, Hatayama et al. 1995, Field 1996, Malagnini et al. 1996, Cornou et al. 2003). In smaller structures short travel times for the laterally propagating surface waves, the complexity in the structure geometry, and the possibility of multiple reverberations between the basin borders result in very complex wave fields that consist of multiple waves mixed in the early S-wave portion (and in the P-wave part as well). Not long since, however, such waves could be observed directly in several small basins: the Caille valley in the French Alps (Gaffet et al. 1998), Colfiorito (Caserta et al. 1998, Rovelli et al. 2001) and Gubbio (Bindi et al. 2009) in central Italy, and Parkway in New Zealand (Chavez-Garcia et al. 1999). In each of these cases, direct observation was made possible through dense array recordings and strong constraints for wave motion patterns due to the depth and geometry of the basin. However, since such dense array recordings are only available for a few basins (and not for Santiago de Chile) there is the need of considering an approach that can deal with the limitations of single station recordings.

Recently, Stockwell et al. (1996) introduced the S transform as a tool for optimal time-frequency analysis of geophysical signals. The transform, which is an extension of the continuous wavelet transform, has already been used by several authors (e.g. Pinnegar and Eaton 2003, Zhu et al. 2003, Goodyear et al. 2004, Schimmel and Gallart, 2005, 2007, Askari and Siahkoohi 2007) for optimally filtering seismograms. One of the main advantages of analyzing seismograms by means of the S transform is that a denoising procedure can be easily coupled to a time-frequency filtering of the signal. An application combining the thresholding denoising method with time-frequency filtering of the seismogram has recently been proposed by Parolai (2009). This technique allows an optimal identification of secondary phases (like dispersive surface waves) which are of particular interest in site effect studies when 2D and 3D effects are amplifying and extending strong ground motion durations.

4.2.3.2. Application to real data

For several of the stronger local and regional events recorded by the network installed in 2010 large long-period phases following the S-waves for stations located in the basin can be observed. As an example Figure 4.3 shows the recording of the local 5 April 2010 event (03:32:14 UTC, M=4.6), occurring around 55 km west of the edge of the basin, and the corresponding normalized S transform.

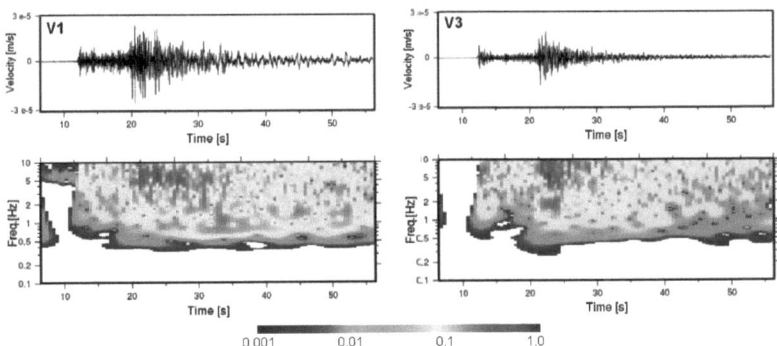

Figure 4.3: Top: The vertical component seismograms of the local M=4.6 earthquake on 5 April 2010, 03:32:14 UTC recorded at soil station V1 (left) and at rock station V3 (right, see Figure 3.1 and Table 3.2). Bottom: Corresponding normalized S transform for V1 (left) and V3 (right).

On the soil site in the basin (station V1, left in Figure 4.3), there is a large phase following the S-wave arrival by about 10 s. In the time segment between 22 and 46 s, the frequency content of ground motion is mainly between 0.4 and 3 Hz and a clear arrival of waves at low frequencies is depicted. On the contrary, this later phase is not observable on the rock site (station V3, right in Figure 4.3). Furthermore, also in the seismogram of station V1 the duration of shaking is significantly longer compared to the rock station. However, higher frequencies mask the arrival on the seismogram, making troublesome, for example, standard polarization analysis.

By combining the denoising of seismograms by means of the S transform and time-frequency filtering (i.e. by multiplying by 0 all undesired spectral components of the S transform and by 1 those that should retain) it is possible to extract the arrival of the dispersive wave train as shown by Parolai (2009). The noise is completely removed and the dispersive wave train excellently imaged (Figure 4.4). In fact, clear low frequency arrivals are followed by higher frequencies. In the pure seismogram, however, it might be difficult to identify this dispersive wave part.

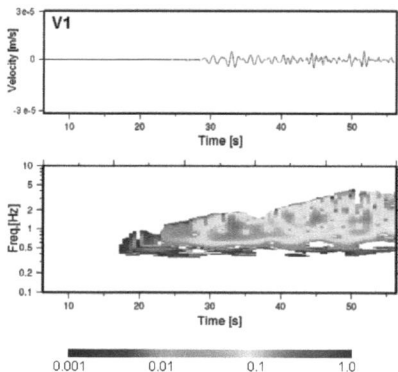

Figure 4.4: Top: The denoised and filtered vertical component seismogram of the local M=4.6 earthquake on 5 April 2010, 03:32:14 UTC at soil station V1. Bottom: Corresponding normalized S transform.

In general, the occurrence of low-frequency dispersive phases following the S-wave is also observed for several other events on soil sites in the basin. Figure 4.5 shows the seismogram and the corresponding normalized S transform for a regional event (M=5.2, latitude 34.65° S, longitude 71.82° W). Again, an arrival of dispersive waves can clearly be identified at around 60 s. Note, however, that the dispersive phase in Figure 4.5 has about the same delay with respect to the S-wave as in Figure 4.3 although the hypocenter for that event is 150 km to the south-west of the basin edge, i.e. around 100 km more from station V1. This indicates these waves must be generated locally, and thus significantly increasing the duration of shaking in the low-frequency range at the basin sites. As will be discussed in chapter 4.7, the resonance frequency of several tall buildings that have been seriously damaged during the 2010 Maule earthquake is expected to overlap with the frequency range of these waves.

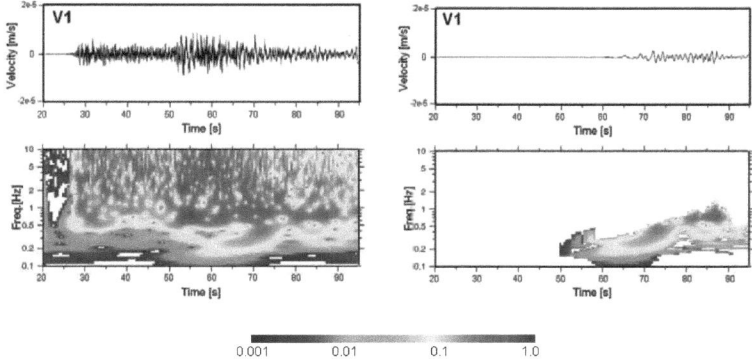

Figure 4.5: Left: The vertical component seismograms of the regional M=5.2 earthquake on 1 April 2010, 12:53:07 UTC recorded at station V1 (top) and the corresponding normalized S transform (bottom). Right: The denoised and filtered vertical component seismogram (top) and the corresponding S transform (bottom).

The wave train for the regional event in Figure 4.5 starts at lower frequencies than the one for the local event in Figure 4.3. As stated before, the local event occurred in the due west of the basin whereas the regional event occurred in south-west direction. One possibility for the different behavior might be the sediment thickness (see Figure 2.4). Here, the variation of dispersion may be attributed to a shift of the frequency bandwidth over which dispersion can occur for the considered soil thickness and velocity contrast between the basin and underlying bedrock (Narayan 2010).

For further characterization of the basin edge generated waves we make use of particle motion analysis which can distinguish between Rayleigh, Love, and body waves. Due to the delay of the dispersive phase with respect to the S-wave there is a high probability that the basin edge generated phases can be identified as surface waves. Although Cotte et al. (2000) showed that long-period (20 s to 40 s) edge generated surface waves can be deviated up to 30 ° relative to the basin edge, and Knopoff et al. (1996) calculated deviations up to 12.5 ° for Rayleigh waves with periods of 15 s, we assume here for the sake of simplicity that for the frequency range considered here (f > 0.2 Hz) these locally generated waves propagate perpendicularly away from the basin edge. Therefore, the horizontal shearing motion from Love waves should be found entirely in the edge-parallel direction while the elliptical particle motion of the Rayleigh wave should be visible entirely in the edge normal and vertical components (Adams et al. 2003,

Yalcinkaya and Alptekin 2005). Figure 4.6 shows the denoised and filtered recordings rotated to these directions.

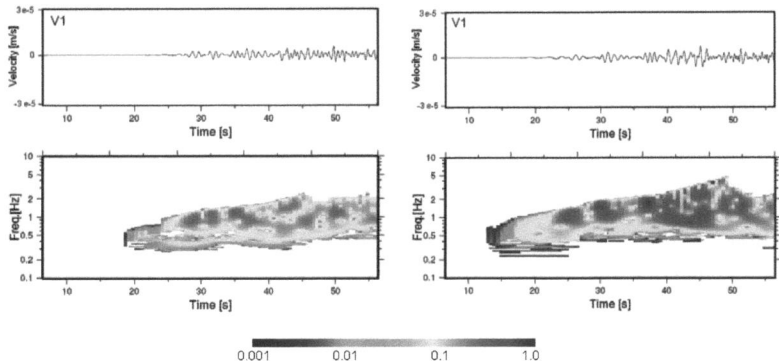

Figure 4.6: Top: The denoised and filtered seismogram of the M=4.6 earthquake on 5 April 2010, 03:32:14 UTC at soil station V1 for basin edge parallel (left) and basin edge normal (right) components. Bottom: Corresponding S transform normalized to the maximum of the two components. The vertical component has already been shown in Figure 4.4.

Both time-frequency representations of the filtered seismograms cover about the same area. However, according to Cornou et al. (2003), basin edge induced Rayleigh waves can carry up to 50 % more energy than Love waves. Similarly, in Figure 4.6 basin edge normal components carry significantly more energy than basin edge parallel components. Therefore, it should be easier to identify properties of Rayleigh waves when using polarization analysis. One has to keep in mind that real signals, although filtered, have highly time-dependent polarization properties. In the time-frequency domain these properties can be represented by the dip and the strike of the polarization ellipse (Pinnegar 2006). Herein, the dip represents the inclination of the ellipse to the horizontal plane ($0 < \text{dip} < \pi$). If the dip is smaller than $\pi/2$, the particle motion is counter-clockwise as viewed from a position having large displacement in the positive z direction; if the dip is larger than $\pi/2$, the particle motion is clockwise. The strike is the azimuth of the ascending node (the point at which the function crosses the horizontal plane in the positive z direction), measured counter-clockwise (see Figure 4.7).

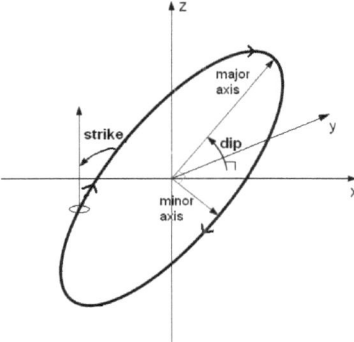

Figure 4.7: Schematic representation of the polarization ellipse with its geometric parameters mentioned in the text.

In general, these spectra are quite difficult to interpret since this technique assigns a value to the strike and to the dip at every point on the time-frequency plane, regardless of whether the ellipse associated with that time and frequency significantly contributes to the signal. One way of addressing this problem can be using two different shades of each color to denote the dip angle, with the brighter shade being used only for these parts in which large amplitudes in the major axis time-frequency representation (i.e. in the S transform) occur, and dimmer shade used everywhere else (Figure 4.8). As one might expect the Rayleigh wave with its vertical ellipse plane turns out to have a dip close to $\pi/2$. This value is reasonably stable over the extent of the Rayleigh wave signature. Other dip values, which can be found in Figure 4.8, might be exhibited by Love (or other) waves and may seem surprising at first since Love wave motion is nearly horizontal, having a dip close to 0. The ellipses that describe Love wave motion are in general extremely elongated. To this regard, one has to keep in mind here that the dip of the ellipse plane is not the same as the plunge of the major axis of the ellipse. Purely linear motion is unlikely to be encountered in real seismograms, since noise contamination will always introduce some ellipticity which could not be removed completely although the seismograms have been filtered.

The strike of the Rayleigh wave (Figure 4.8 bottom) shows a quite stable behavior with values near 0. This means that the particle motion is displaced towards the positive radial direction when it rises through the horizontal plane which is consistent with the retrograde motion of the Rayleigh wave. For the transverse Love wave, one would expect strike values near ± π/2 relative to the positive radial direction. The reason for two possible values is that the strike is defined to be the azimuth that the particle has when it crosses the horizontal plane in the direction of increasing z. Thus, changing the strike by ± π is equivalent to reversing the direction of particle motion. This might explain the light red and light blue parts in the time-frequency representation of the strike.

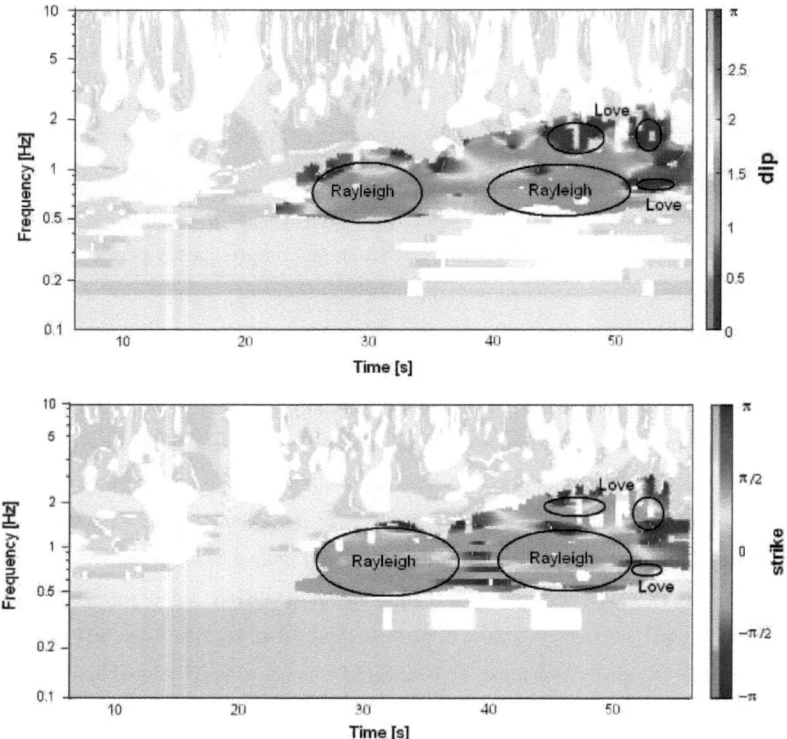

Figure 4.8: Top: Time-frequency representation of the dip of the elliptical motion. The Rayleigh signature exhibits a classic vertical behavior with a dip near π/2. Horizontally polarized Love waves show a dip close to 0. Bottom: Time-frequency representation of the strike of the elliptical motion. 0 represents the positive radial direction. This shows the alignment of the Rayleigh ellipse with the propagation direction (retrograde motion). Strike values near ± π/2 indicate the transverse orientation of the Love phase.

Thus, although the signals had been filtered even small noise contributions will change the pure linear motion. Since the influence of noise each with strike and dip is different, this might also explain why the polarization properties of Rayleigh and Love waves do not cover the same area in the time-frequency representations, respectively. However, the polarization properties of surface waves can clearly be identified. In conjunction with their respective delay regarding the S-wave arrival this indicates the existence of basin edge induced surface waves in the Santiago basin causing longer shaking duration at several sites. However, to draw a general conclusion our azimuthal coverage is not the best since all recorded events occurred to the west and south-west of the basin. We expect the excitation of surface waves to be different for other events (depending, e.g., on the azimuth of the incoming seismic waves), although we are quite sure that one would also identify the eastern edge of the basin as a diffractor, too. However, we did not observe an excitation of surface waves for all analyzed events; on the other hand, no statistical dependency on the distance of the events could be found at first glance.

Therefore, further investigations are necessary to clarify the possible influence of magnitude and back-azimuth on the generation of surface waves. Besides, our previous investigation is only focused on the local basin structure. As outlined by Gaffet et al. (1998), regional heterogeneities also can contribute to reload the basin with seismic energy. Since there is major lack of this information, additional studies are necessary to further quantify the occurrence of these waves.

Yet, without having a detailed knowledge thereof a further analysis of ground motion based on the spectral ratios of the earthquake can be carried out.

4.2.4. Frequency domain analysis

To calculate the spectra for all the events, the recordings were first checked visually and only those showing good signal-to-noise ratio allowing for detecting P- and S-wave arrival were considered; then time windows for P-wave and S-wave analysis were selected. The P-wave window starts 0.7 s before the P-wave arrival and ends before the S-wave arrival, when the P-wave energy reaches 90 % of its maximum. The same applies for the S-wave window. After correcting for instrumental response, each window was cosine tapered (5 %) and a Fast Fourier Transformation (FFT) for each seismometer component was performed. Spectral amplitudes were smoothed using the Konno and Ohmachi (1998) recording window (b = 40), ensuring smoothing of numerical instabilities while preserving the major features of the earthquake spectra that were considered valid only when the signal-to-noise (S/N) ratio is greater than 3. The horizontal component spectra were calculated, considering the root-mean square average of the NS and EW component spectra. First, the EHV ratio was calculated for all events separately for P- and S-wave windows (Figure 4.9).

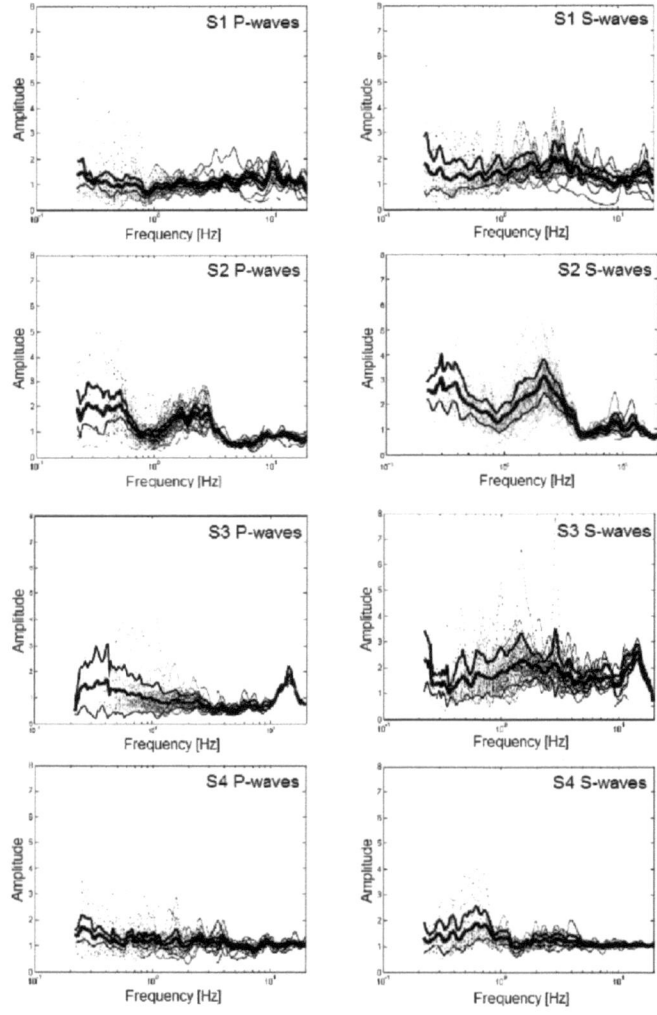

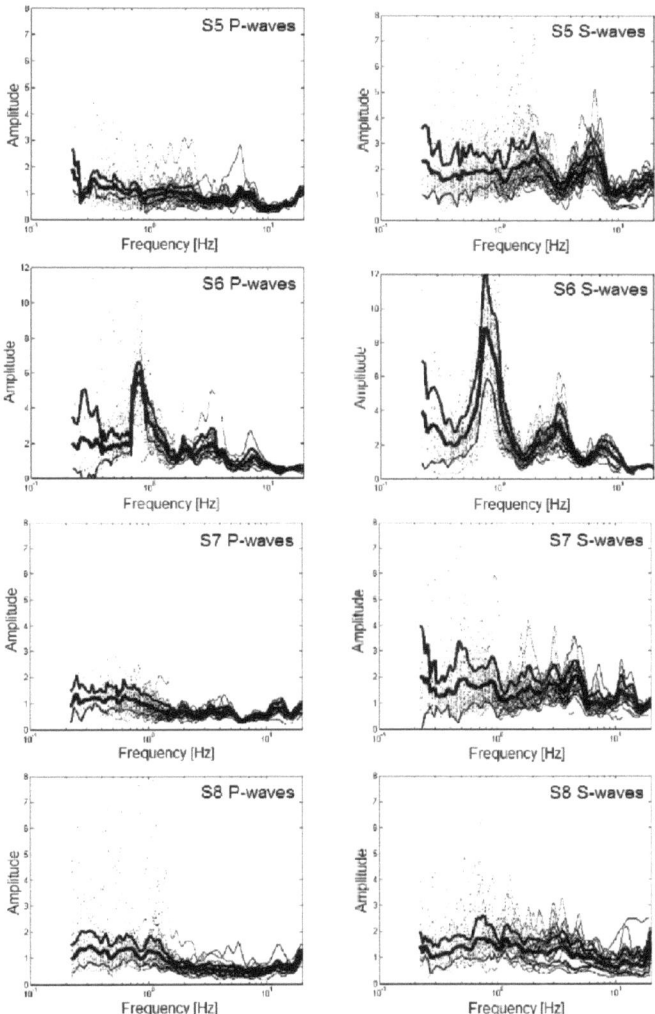

Figure 4.9: EHV calculated for P- and S-waves for all events with a signal-to-noise ratio larger than 3. Thick lines show medium plus / minus one standard deviation. Note the different amplitude scaling for station S6.

It has been shown that EHV and SSR methods usually provide site responses with similar shapes when only the S-wave part of the seismogram is used; the P-wave part is found to provide consistent results only in some cases (see e.g. Castro and the

RESNOM working group 1998). Parolai et al. (2004b) found that EHV P- and S-waves show the same features, although the amplitude of EHV P-waves is smaller. Therefore, we considered both parts separately. Additionally, the SSR method was applied to all recordings using S1 as a reference station due to its location on igneous rock and its nearly flat response in the EHV spectrum. SSR curves for all stations are shown in Figure 4.10. The shape of site responses between P- and S-waves is again quite similar.

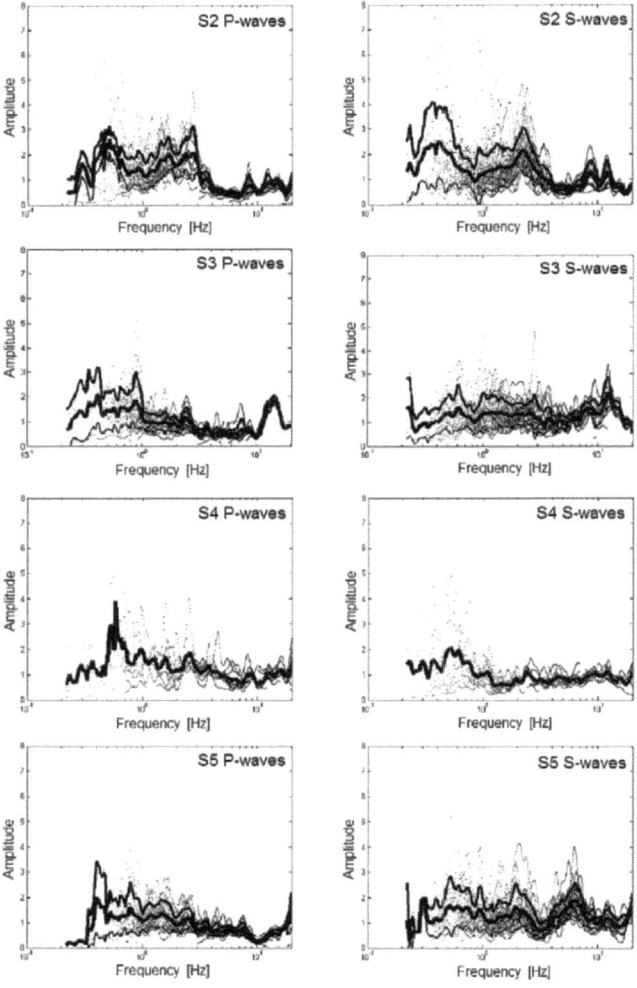

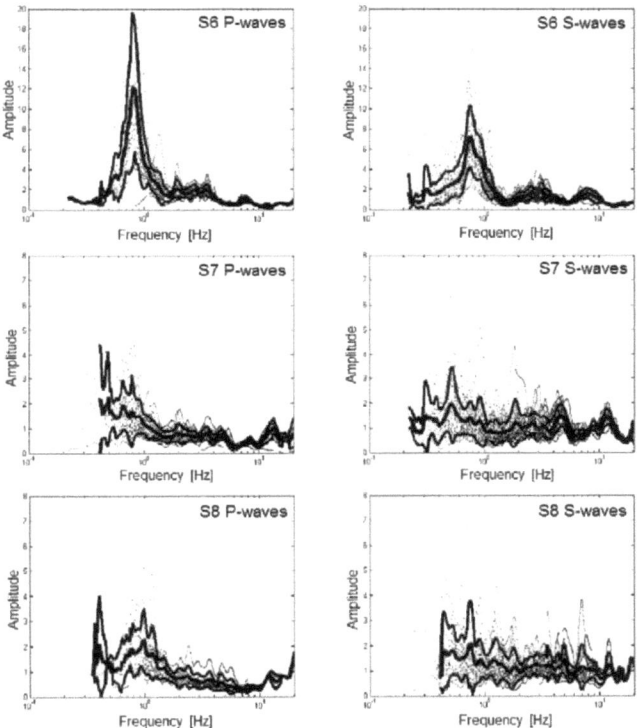

Figure 4.10: SSR calculated for P- and S-waves. Thick lines show medium plus / minus one standard deviation. Note the different amplitude scaling for station S6. For station S4 no standard deviation has been calculated due to too few recordings.

4.3. H/V ratio of ambient noise: Data acquisition and analysis

As all the stations of the network in 2008 were set to continuous recording, a huge amount of ambient seismic noise was recorded, allowing further investigations regarding the long term stability of the NHV peaks and the verification of the existence of differences in the results, when using recordings from day and night time. The data also provided the means for comparing amplification factors obtained from the NHV peak with those determined by other methods.

For data processing each noise recording was divided into 60 s-long windows tapered with a 5 % cosine function. As a rule of thumb, it is generally accepted (Parolai et al.

2001, Bard and SESAME WP02 team 2005, Picozzi et al. 2005) that the shortest window length has to be selected in such a way as to include at least ten cycles of the lowest frequency analyzed. So, a window length of 60 s can be considered long enough to accomplish the analysis down to a frequency of at least 0.2 Hz. Instrumental correction and smoothening were carried out in analogy to earthquake data analysis. After checking visually for anomalies between Fourier spectra of the NS and the EW components, both spectra were then averaged (root-mean square) obtaining the horizontal component Fourier spectrum. Afterwards we calculated the spectral ratios between the horizontal and vertical components, and finally we determined the logarithmic mean of all the H/V ratios for a given site. For our studies, the number of selected windows (20 to 30) for this frequency range guarantees stable results, following Picozzi et al. (2005).

In order to estimate the stability of the peak in the NHV curve of the installed seismic stations both in terms of frequency and amplitude, the spectra were calculated using data recorded some time after the installation of the stations and before their removal with a time lag of seven weeks. For both dates and all eight stations, NHV curves were calculated at midnight (04:00 UTC) and at noon local time (16:00 UTC) on the same day, respectively. Results are presented in Figure 4.11. The analysis clearly shows that the shape of the NHV curves is not identical even though no large differences between recordings at midnight and at noon seem to occur.

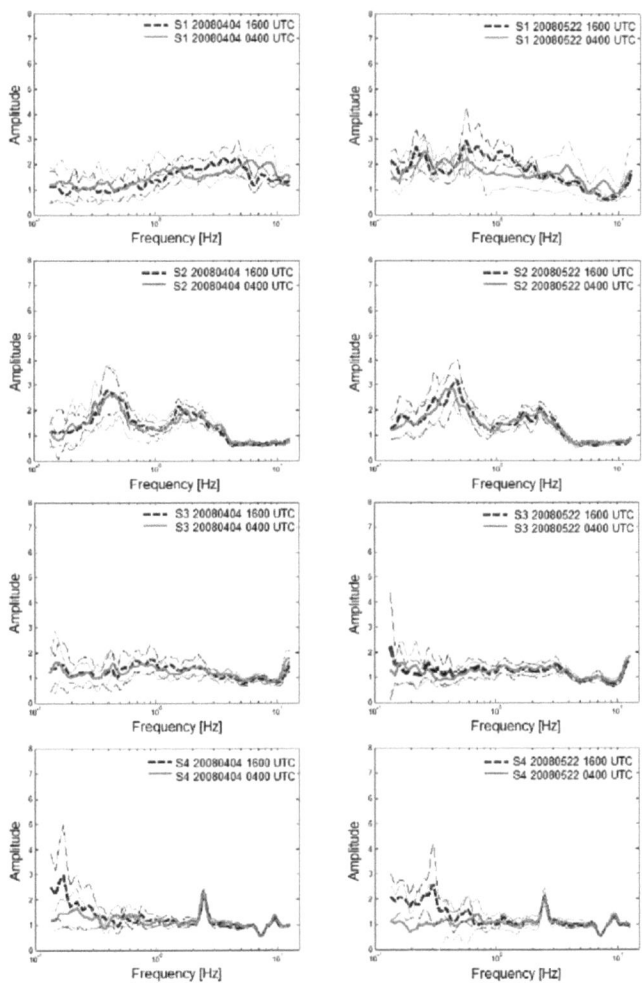

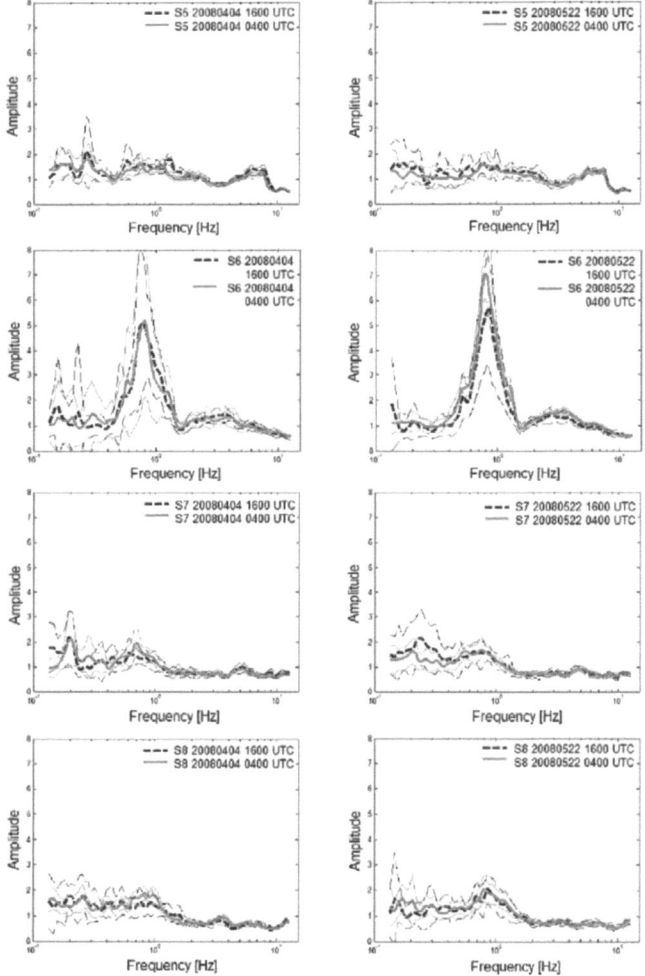

Figure 4.11: Average NHV plus / minus one standard deviation calculated at noon (16:00 UTC) and at midnight local time (04:00 UTC) for the same day. Figures on the left show recordings of 4 April 2008, figures on the right recordings of 22 May 2008.

An analysis comparing Figures 4.9 (EHV curves), Figures 4.10 (SSR curves), and Figures 4.11 (NHV curves) is outlined below.

4.4. Results and discussion

4.4.1. Seismic event and microtremor recordings: comparison of different techniques

- S1 (reference station, no SSR curve)

As expected, no clear peak can be identified in the NHV curve although the station shows little amplification over a broader frequency range (< 5 Hz) in the NHV spectral ratios. To check, if topographic effects are responsible for the small amplification we investigated the directional variation of the NHV spectral ratio according to Del Gaudio et al. (2008). Therefore, we calculated an average NHV spectral ratio consisting of 20 NHV curves selected randomly using recordings from day and night time over the entire installation period. In Figure 4.12 the mean NHV values are represented by a polar diagram. Although differences between the maximum and minimum amplitude at the same frequency are small, a clear preferential direction in the azimuthal distribution of the amplitude is shown: the relative maximum is oriented along N 70° E, the maximum slope direction, confirming the results found by Del Gaudio et al. (2008). However, all outcrops of the igneous bedrock show significant topographic slopes; hence, this effect might also be observed at any other site on the bedrock close to the city.

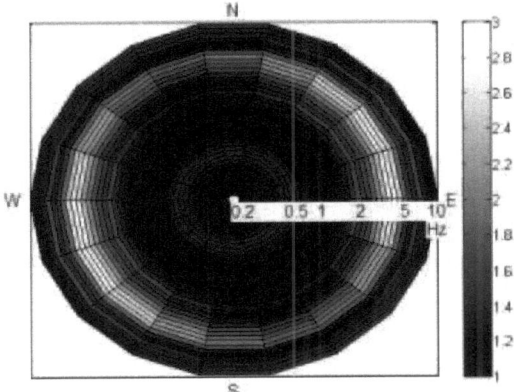

Figure 4.12: Polar diagram of mean NHV spectral ratio for station S1 calculated at 20° azimuth intervals for an average of 20 randomly selected NHV measurements.

For the NHV curves recorded on 22 May 2008, there are small differences between the recordings at noon and at midnight (Figure 4.11). Due to limited sites on the San Cristóbal hill providing a power supply, the installation of the station was not far away from the mountain station of the funicular onto the hill. As this area is much frequented by people during daytime, the difference might be explained by these disturbances.

When looking at the earthquake data, the EHV curve of the S-waves shows a slightly higher level of amplification with respect to both the P-wave windows and to the NHV curve. Although only a slight influence of the topography on the amplification for NHV spectral ratios at S1 is found, this trend cannot be seen when considering EHV spectral ratios: no azimuthal distribution is found. So the lack of significant amplification might justify to choose this station as a reference.

- S2

Although all the NHV, EHV, and SSR curves show a consistent shape, the amplification of the S-wave windows is generally slightly larger compared to the P-wave windows. Whereas the second peak around 2 Hz is smaller than the first one at 0.45 Hz in the NHV curve, both peaks are of the same amplitude for EHV and SSR. The secondary peak at around 2 Hz can be explained by considering local geological conditions: as described in chapter 3.1.1, at this site a layer of pumacit is overlaying gravel, therefore a second impedance contrast might reasonably explain this peak. The frequencies at which the peaks occur seem to be at a first order consistent with the thickness of the sediments obtained by the gravimetric data set. Additionally, some events show (quite small) peaks in the EHV and SSR curves around 8.7 Hz and 13 Hz, which the NHV method is unable to detect. These peaks are likely to be related to relatively superficial impedance contrasts which was already suggested by Bravo (1992) and Rebolledo et al. (2006) in their geological studies of the pumacit layer. No differences between the recordings made at day and night time are visible in the NHV curves.

- S3

Although there is a strong scattering, no striking significant peak can be seen in the noise or in the earthquake curves. Several small peaks for some events appear in the

EHV curves. In general, the amplitude of the S-wave window curve is slightly larger than the one of the P-waves.

- S4

While the P-wave curve of the EHV remains flat, the S-wave curves of EHV and SSR show some slight amplification in the frequency range between 0.5 and 0.9 Hz. Due to the few recorded earthquakes with good signal-to-noise ratios, no statements can definitively be made about the P-wave windows of the SSR results. On the other hand, although the NHV curves seem to be flat, a clear peak at exactly 2.5 Hz and a small one at 9.3 Hz are visible in all NHV curves, also slightly indicated in the SSR curves. When calculating the spectra for each component for the microtremor recordings separately (Figure 4.13), obvious differences occur.

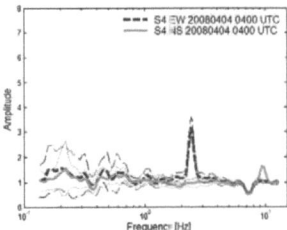

Figure 4.13: EW/V and NS/V spectra plus / minus one standard deviation for the recording on 4 April 2008 of station S4.

In fact, the peak at 2.5 Hz is not found in the (NS)/V spectral ratio of noise while it can clearly be seen in the (EW)/V spectral ratio, with the opposite situation for the small peak at 9.3 Hz. The standard deviations for all curves remain quite small. As it is really unlikely to have two sources oriented at exactly 90 degrees from each other without any influence on the orthogonal component and since it has already been reported (e.g. Gallipoli et al. 2004b, Cornou et al. 2004) that noise and even earthquake recordings are strongly influenced by the proximity of structures, we believe that these frequencies might be related to the first two translational modes of the building in which the sensor was located. Furthermore, the fact that these peaks are not related to the soil response is also supported by the results of the single stations NHV measurements determining much lower frequencies for this area (see chapter 4.6 and Figure 4.16).

This example clearly confirms that all components of the Fourier spectra have to be checked carefully, and effects of the structure where the sensor is installed have to be accounted before analyzing the results, so as to avoid misinterpretation.

- S5

Although the curves have similar trends, the amplification retrieved by the S-wave window curves are higher than those of the P-wave windows, showing a peak at 6 Hz which is only foreshadowed in the NHV curve and can hardly be seen in the P-wave curves. Over a broad frequency band between 0.5 Hz and 3 Hz, all curves show slight amplification with amplitudes marginally below 2.

- S6

All curves show a clear peak at 0.9 Hz with amplitudes ranging between 5 (NHV curve) and 12 (SSR P-waves). The peak of the fundamental resonance frequency is as a first approximation consistent with the thickness of the sedimentary cover provided by interpreting the gravimetric data. Peaks of higher harmonics at around 3 Hz and 8 Hz can also be observed in the earthquake data, especially on the S-wave window curves, while the NHV curve show only small bumps over that frequency range.

- S7

Due to a low signal-to-noise ratio, no reliable SSR curve below 0.4 Hz could be determined. All curves remain quite flat. Nonetheless, when checking the EHV and SSR S-wave curves, a few events show amplification at 4.8 Hz and 13 Hz, respectively, but the amplitudes remain rather small. This trend is not visible for both P-wave curves. As the translational modes of the building are excited much better by S-waves than by P-waves (e.g. Wegner et al. 2005) and as the major axes of the building almost coincide with the NS and EW direction both small peaks might again be related to the first two translational modes of the building. In the NHV curves, no clear peaks can be found, although there are slight elevations for frequencies around 0.2 Hz and 0.7 Hz. Considering the thickness of the sediments obtained by gravimetric data, the fundamental frequency is expected to occur around 0.7 Hz, i.e. the first peak is not the fundamental one. In addition, the small amplitude of the peaks might indicate a low impedance contrast.

- S8

Due to a low signal-to-noise ratio, no reliable SSR curve below 0.4 Hz could be determined. Large scattering appears with some events showing amplification around 1 Hz, which might also be supported by a bump in the NHV curve recorded on 22 May, while this bump cannot easily be seen in the curves of 4 April. Considering a sedimentary cover thickness of around 95 m amplification at around 1 Hz seems to be realistic.

4.4.2. Summary

Generally, there is a good agreement in the shape and the location of the fundamental resonance frequency found in the SSR, EHV, and NHV curves. EHV and SSR spectral ratios show similar shape and amplitude, because the small amplification found at station S1 due to topographic effects does not influence the results at the other stations. As only events with an epicentral distance of five times the maximum interstation distance have been considered for analysis, the SSR technique is found to provide a more reliable estimate of the site response, because effects due to the path and the radiation of the source are minimized, respectively. Moreover, although the recorded earthquakes do not cover the entire back azimuth range we did not observe any dependence of the EHV and SSR amplitudes on earthquake location, in consistency to e.g. Bonnamassa and Vidale (1991).

In general, the level of amplification is surprisingly low at several stations located within the basin (i.e. S4, S5, and S8). Reliable EHV and SSR site responses can usually only be obtained, when earthquakes are distributed all around the stations at different distances. Due to the tectonic situation, the recorded earthquakes are mainly clustered in two narrow azimuthal ranges, as can be seen in Figure 4.2, but no influence is believed to arise from this shortcoming due to the relatively large number of events used. Therefore, the small amplitude at several sites might give a first hint at a low impedance contrast, a conclusion supported when considering the soil conditions below the sites (see site characterization above) and also by the findings of Bravo (1992).

Differences between NHV and earthquake spectral ratios can be found in terms of the amplification level: most of the earthquake data show a higher level of amplification than NHV curves for large ranges of the frequency band. This is also strengthened by Figure 4.14 where the ratio of SSR divided by NHV for all stations in the frequency range 0.2 Hz to 20 Hz (0.4 to 20 Hz for S7 and S8) is shown.

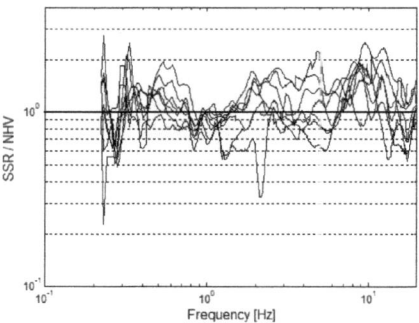

Figure 4.14: Ratio between standard spectral ratio and H/V site responses from seismic noise.

NHV seems to give a lower amplification, especially for lower and higher frequencies, whereas for frequencies around 1 Hz the ratio between SSR and NHV takes a value below one. Some stations (e.g. S2 and S6), located on thick sediments, have clear peaks related to higher harmonics in the EHV and SSR curves with amplifications sometimes exceeding 3. Because NHV is generally expected to estimate only the fundamental resonance frequency (or to provide amplifications over a narrow frequency band around it), it could be expected to underestimate more significantly the level of amplification at frequencies higher than the fundamental one. Therefore, the ratio between SSR and NHV is expected to be frequency dependent and to differ between analyzed stations, depending on the position of the peak relevant to the fundamental frequency.

Additionally, when comparing spectral ratios of P- and S-wave windows, both spectra share similar features as long as the frequencies of maximum amplification are considered (if appearing in the P-wave curve), but amplitudes in the P-wave curves are often considerably lower. However, this difference might be explained by the hypothesis that the P-wave window results are mainly related to converted waves. Since a low impedance contrast seems to exist, the amount of conversion is limited and the signal-to-noise ratio in the windows is low.

Finally, in Figure 4.11 it can be seen that the shape of all NHV curves is quite stable, which means that if there are any peaks, they occur at almost the same frequencies and show the same amplitudes. Also, no differences between recordings during daytime or nighttime are observed, with the exception of station S1 on 22 May 2008, as discussed above. Since transient noise is expected to be generated mainly by nearby sources and generally affects noise spectra at frequencies higher than 1-2 Hz (e.g. McNamara 2004), this might confirm that transients have no (or only a little) influence on the NHV ratio, a result already shown by several studies (e.g. Mucciarelli et al. 2003, Parolai and Galiana-Merino 2006).

We can therefore conclude that no noticeable NHV variation with time can be observed. This result might be used for further experiments by analysing noise measurements within the city, as the amplitude and the fundamental frequency of the NHV peak is not influenced by the time of measurement and the environmental conditions.

4.5. Single station NHV measurements

As mentioned, in addition to the earthquake recordings, measurements of seismic noise were carried out at 146 different sites in the northern part of Santiago de Chile (measurement sites are shown in Figure 3.1). There exist several criteria how to qualify the reliability of a peak, including its absolute amplitude, the relative amplitude value with respect to other peaks, and by considering the standard deviation. We used the following guidelines proposed by the SESAME consortium (see Bard and SESAME WP02 team 2005 for further details):

- The maximum peak amplitude of the NHV curve is higher than 2.
- There is a lower threshold frequency f^- in the frequency range $f|_{H/V}/4 < f^- < f|_{H/V}$ such that the amplitude ratio is fulfilling $A|_{H/V} / A|_{f^-} > 2$ (Therein, $f|_{H/V}$ indicates the fundamental frequency and $A|_{H/V}$ the amplitude at that frequency).
- There is an upper threshold frequency f^+ in the frequency range $f|_{H/V} < f^+ < 4 \cdot f|_{H/V}$ such that the amplitude ratio is fulfilling $A|_{H/V} / A|_{f^+} > 2$.
- The standard deviation has to be lower than a frequency dependent threshold.

To be sure that the site is likely to experience ground motion amplification, at least three of these four criteria should be achieved; then the peak frequency can be considered as a reliable estimate of the fundamental frequency of the site.

All average NHV curves were systematically analyzed following the SESAME recommendations. In the end, 115 reliable NHV curves fulfilled these criteria. In another 16 curves, a clear peak was visible, although not fulfilling at least three of the four criteria mentioned above, while for 15 curves, no reliable peak could be determined. Hence, these 15 latter curves were not considered for further analysis. In Figure 4.15, examples of the different kinds of curves are shown.

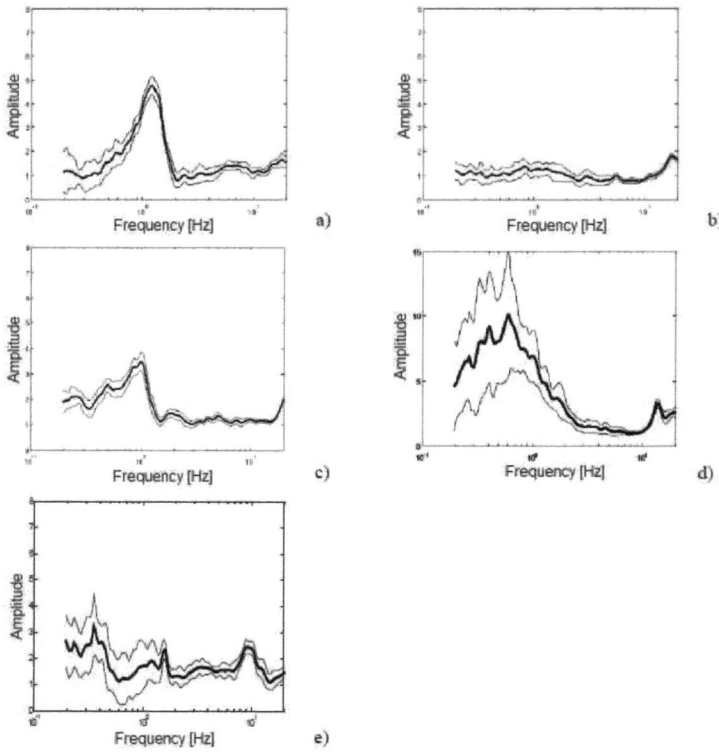

Figure 4.15: Examples for different NHV spectral ratios plus / minus one standard deviation. Please note the different amplitude scaling for Figure d. See text for further discussion.

When the NHV peak is clear (Figure 4.15 a), a large impedance (i.e. velocity) contrast between the sedimentary cover and the bedrock is expected to exist, hence it is very likely that ground motion is amplified at this site. Flat NHV curves (Figure 4.15 b) and peaks with low amplitudes are located at the surface where the sediments are characterized by high density values, i.e. mainly gravel. A low impedance contrast resulting from stiff sediments underlain by bedrock is likely to exist and therefore one might expect rather small amplification. Although flat NHV curves within a much wider area in the Santiago basin were found by Bonnefoy-Claudet et al. (2009), these authors suggest that NHV curves showing no peak are from stiff sediment sites. Curves like those in Figure 4.15 c show quite broad peaks spread over the entire basin, with no correlation with topographic characteristics and local geology. The influence of wind can be excluded as the sensor and cables were protected and no measurements were made in the immediate vicinity of trees or buildings. On the other hand, gravimetric data (Araneda et al. 2000) and cross sections (Pasten 2007) show that at some sites, the thickness of the sedimentary cover varies over short distances (see Figure 2.4). Previous studies (Wooleroy and Street 2002) show that broad NHV peaks are usually obtained at sites with a complicated subsurface geometry and large near-surface shear wave velocity contrasts. Although there might sometimes be difficulties in determining an exact fundamental frequency, at least a bound can be identified where one can be confident that amplification will occur. Large standard deviations like those shown in Figure 4.15 d often mean that ambient vibrations are strongly non-stationary and undergo some kind of perturbations that might be removed by a longer recording duration. Nonetheless, as shown here, a clear peak is visible which can be used for determining the fundamental resonance frequency of the site. Finally, curves such as shown in Figure 4.15 e would actually have to be discarded, as they do not fulfill at least three of the four standard criteria, here a too marginal decrease for low frequencies and a large standard deviation. Nonetheless, three clear peaks at around 0.35 Hz, 1.7 Hz, and 10 Hz are visible. (Although the peak at 1.7 Hz is quite narrow, the individual spectra suggest that it is not generated by industrial noise.) As the recording presented in Figure 4.15 e was only 900 m away from the site where station S2 had been installed and also the shapes of Figure 4.15 e and the EHV and NHV curves of S2 (Figures 4.9 and 4.11) look quite similar, the peaks are considered to be reliable due to their spatial coherency.

Altogether, although 31 curves do not always fulfill all the standard conditions for a reliable curve as mentioned above, 16 of these were used for further analyses because a peak was clearly visible and the criteria hereunto seemed to be too restrictive. Most of these measurements were carried out at sites with a low impedance contrast in the eastern part of the investigated area, which means that the peak amplitude is often slightly below 2 and the decrease (predominantly towards lower frequencies) is too small, while neighbouring sites fulfilled the criteria showing peak amplitudes beyond 2. Usually, the foremost measurements would not be considered for further analyses, but their reliability (similar to the example of Figure 4.15) was assessed due to spatial coherency and geological and geophysical data sets. Hence, because of the experimental character of the NHV method, visual inspection of the data should always be performed.

4.6. Fundamental resonance frequency map of the investigated area

The analysis of the fundamental resonance frequencies calculated for all sites allowed us to draw a map of the fundamental frequency of the investigated area (Figure 4.16).

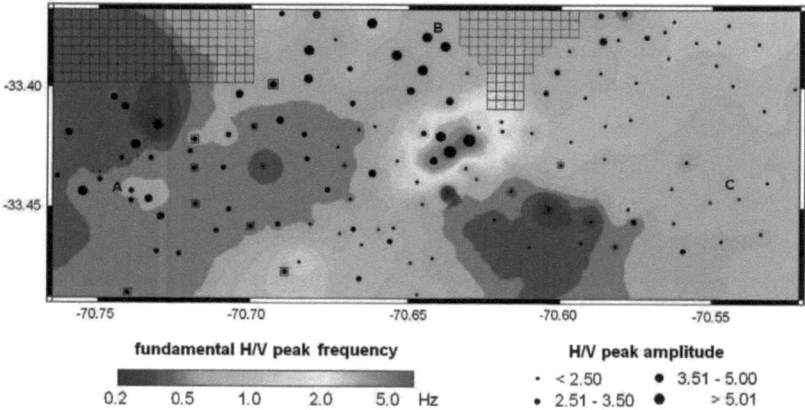

Figure 4.16: Map of the fundamental resonance frequency of the investigated area. Circles indicate microtremor measurement sites that have been considered for interpolation. Spots enclosed by squares represent broad H/V peaks (see text for further discussion). Diameter of the spots corresponds to H/V peak amplitude. In the hatched area, no measurements were carried out; these values have not been considered for mapping the fundamental frequency hence results are only due to interpolation. Letters indicate sites where pictures shown in Figure 4.17 have been taken.

In general, the fundamental resonance frequency varies slightly in space without large differences between neighbouring sites. Almost the entire area is characterized by relatively low frequencies below 1 Hz. Only when one considers sites close to the city center and around the San Cristóbal hill, the fundamental frequency does increase more rapidly. In this part of the basin, the thickness of the sedimentary cover is decreasing and the bedrock outcrops, as evident by the San Cristóbal hill (see Figures 2.3 and 2.4). Additionally, there is also a rapid decrease in the sediment thickness towards the northwest of the investigated area, but no increase in the fundamental frequency can be observed there because no noise measurements have been carried out in the immediate vicinity of the Renca hill. Only two nearby sites show a slight trend towards higher frequencies.

Recordings corresponding to clear and narrow peaks are in great majority located in the northern and western parts of the investigated area. This is a first hint of the existence of a strong velocity contrast at depth between unconsolidated sediments and the bedrock. On the contrary, flat H/V curves and peaks with low amplitude are mainly found in the southern and eastern parts of the basin where coarser sediments (mainly gravel) outcrop, consistent with Bonnefoy-Claudet et al. (2009). There, a low impedance contrast between the sediments and the bedrock is likely to exist.

4.7. Correlation between fundamental frequency and damage distribution of the 1985 Valparaiso and 2010 Maule events

Considering that, as a first approximation for Chilean structures, the relationship between the height of a building and its fundamental period of vibration T can be expressed by

$$T \text{ [s]} = (\text{height of structure [m]}) / C, \qquad (1)$$

whereas 50 m/s $\leq C \leq$ 60 m/s for normal structures and 30 m/s $\leq C \leq$ 40 m/s for flexible structures (Guendelman 2000), we expect that for most parts of the city, the natural frequency of the soil will match the frequency of buildings with more than ten storeys, i.e. a height of more than 30 m.

However, during the 1985 Valparaiso and 2010 Maule earthquakes, there was considerable structural damage to small-sized adobe structures located predominantly in the western part of the city. Also few masonry buildings with steel frame show moderate to substantial damage, an expected behavior (Moroni et al. 2004). The observed damage there seems to be predominantly caused by the old aged building stock characterized by a high vulnerability and not due to an overlap of the fundamental frequencies (Figure 4.17 top). One must, however, remember that the NHV method provides only a lower bound of the amplification level and of the frequency, as stated already by Lebrun et al. (2001). Therefore, also higher harmonics might correspond to the eigenfrequencies of these buildings or even to higher harmonics of the buildings. Furthermore, two overpasses on the highway, built only a few years ago, collapsed in that area, two other show minor structural damage (Bray and Frost 2010). Altogether, this indicates the EMS intensity to be VIII (Lars Abrahamczyk, personal communication).

Figure 4.17: Examples of different building stock in the city of Santiago de Chile. Exact locations of the sites are indicated in Figure 4.16. Top (A): Small sized residential buildings in Pudahuel. Middle (B): Commercials building in south-eastern part of Huechuraba with six floors (left) and five floors (right). Both show structural damage (Bray and Frost 2010). Bottom (C): Damaged high rise structures surrounded by undamaged small building stock. Pictures were taken between 11 and 18 March 2010. Please see text for further discussion.

On the contrary, in the business district in the central northern part of the investigated area (south-eastern part of Huechuraba, see Figure 2.2) many (in general modern reinforced concrete) buildings have been partially and seven of them structurally damaged during the 2010 Maule earthquake (Figure 4.17 middle). However, no total collapse was documented for this site (Bray and Frost 2010). Considering that, although these structures are new (less than ten years) and should have been constructed according to actual building codes, most of them have five to ten floors. Since the fundamental resonance frequency of the soil in this area is around 0.7 to 1 Hz (Figure 4.16) the high damage might be due to an overlap of the resonance frequencies of the soil and the present structures. Moreover, since highest NHV amplitude values are found in that area, also the ground motion might have been relatively strong; numerical simulations serve as a further proof therefore, as will be discussed in the following (see chapter 6.3.3). We estimate an EMS intensity of VII to VIII for that area.

In the far eastern districts of the study area modern residential structures up to four floors can be found. Generally, no damage is observed for these buildings; also NHV peak amplitudes are low. In contrast, five tenement blocks with 17 floors each (Figure 4.17 bottom) constitute an exception. These buildings show structural damage, and two towers have been tilted (Daniel Hernández, personal communication).

The reason for this behavior is not surprising since we determined the fundamental resonance frequency of the building by measurements on the top of one tower to be around 0.7 Hz (Figure 4.18 top right). This frequency exactly coincides with the fundamental resonance frequency of the soil for a measurement site around 500 m away (Figure 4.18 bottom right). It seems reasonable to assume that resonance effects are responsible for the damage distribution in that area. Moreover, also higher harmonics of the building can easily be seen in the Fourier spectra. The spectra of the two perpendicular horizontal directions are relatively close to each other showing that there is not much difference between the stiffness in the two directions, an expected result since the floor plan is almost quadrate. However, this might also indicate the first rotational mode of the building; heavy structural subjected to seismic motion, as observed here, is frequently due to torsional effects (Rocco Ditommaso, personal communication).

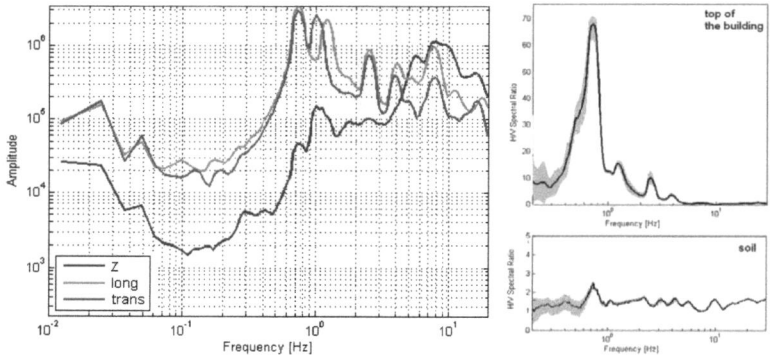

Figure 4.18: Left: Fourier spectra measured on the top of an apartment block which is shown at the bottom of Figure 4.17. Location of the buildings is indicated with letter C in Figure 4.16. Right: Corresponding H/V spectral ratio (top) and H/V spectral ratio of the soil measured around 500 m apart (bottom). Please note the different amplitude scaling for the H/V spectral ratios.

Although the distribution of the fundamental resonance frequency of the soil cannot by far explain the observed damage distribution in the city during the 2010 Maule earthquake, the presence of some eye-catching peculiarities has led to suspect a contribution of soil structure resonance effects. However, to obtain a comprehensive and differentiated overview a further characterization of the soil parameters, which will provide complementary information about local site conditions, has to be taken into account.

Chapter 5

3D shear wave velocity model

5.1. Introduction

Since the 1985 Valparaiso and the 2010 Maule earthquakes clearly have shown that varying soil properties in the Santiago basin have influence on the level of ground shaking and the damage distribution, the microtremor measurements can be used to further characterize local site conditions. In general, two main types of ground motion amplification effects can be distinguished: stratigraphic effects due to the velocity contrast between the sedimentary soil layers on the one hand (Bard and Bouchon 1985, Sánchez-Sesma and Luzon 1995, Bielak et al. 1999, Chávez-García et al. 1999) and topographic effects due to focusing and / or scattering effects around crests and hills (Bouchon 1973, Paolucci 2002, Semblat et al. 2002). Hence, the seismic response might be influenced both by the local geology, which often leaves significant imprint on seismic motion by amplifying the amplitudes of seismic waves and increasing the shaking duration during earthquakes, and the shape of the 3D basin, which may cause the seismic response to differ significantly from that of a 1D layer (e.g. trapped surface waves, see chapter 4.2.3.2). The S-wave velocity structure of unconsolidated sediments down to the bedrock, the impedance contrast between these sediments and the bedrock, as well as the shape of the sediment-bedrock interface, can be regarded as the main controlling parameters of such an imprint. Obviously, an exact identification of these parameters is an important step in seismic hazard assessment.

In order to obtain such detailed models, active in-situ measurements such as shear wave seismics and down-hole and cross-hole techniques can be performed. However, especially in urban areas, it might be difficult to apply these techniques due to the impossibility of making use of explosive sources. For this reason, non-invasive and cost effective passive seismic techniques with single stations or arrays have recently become an attractive option for seismic site effect studies, providing reliable information about the subsurface with a good lateral coverage. Especially over the last decade, H/V spectral ratios and 2D micro-array recordings for estimating surface wave dispersion curves were found to provide very promising results. Based on Nogoshi and Igarashi (1970, 1971), Tokimatsu and Miyadera (1992) found that the variation of microtremor H/V ratios with frequency corresponds to that of the Rayleigh wave for the S-wave velocity profile at the site. Based on a theory for surface (both Rayleigh and Love) waves proposed by Harkrider (1964), Arai and Tokimatsu (2000) presented theoretical formulas for simulating microtremor H/V spectra in which the effects of fundamental and higher modes can be considered. They found that the theoretical H/V spectrum computed for the S-wave velocity profile at a site can closely match the observed microtremor H/V spectrum and, vice versa, that the S-wave velocity structure at a site can be estimated from the inversion of the microtremor H/V spectrum. Several studies have exploited the availability of measurements of ambient noise to derive vertical shear wave velocity profiles (e.g. Fäh et al. 2001, 2003, 2006, Scherbaum et al. 2003, Arai and Tokimatsu 2004, Parolai et al. 2006).

Therefore, the data of the microtremor measurements outlined above are used along with available geological and geophysical data for deriving information about the S-wave velocity structure. After providing a description about how the S-wave velocity model was derived, a detailed characterization of the model will be provided, including a comparison with the results of other available data sets. Finally, it will be tested if the slope of topography may already provide a reliable proxy for the evaluation of v_s^{30} and if it can be appropriate for site response estimation.

5.2. Inversion of H/V ratios for deriving S-wave velocity profiles

It has been shown (Fäh et al. 2001) that the shape of H/V ratios around the fundamental resonance frequency and around the first minimum of the average H/V ratio depends mostly on the layering of the sediments. Therefore we inverted each H/V ratio curve individually under the assumption of a horizontally layered one-dimensional structure below the site. The synthetic H/V ratios, calculated during the inversion procedure, are suitable of allowing the effect of Rayleigh and Love waves (also higher modes) to be taken into account, following Fäh et al. (2001). A genetic algorithm forms the basis for the inversion scheme due to the non-linear nature of the problem. The inversion was performed by means of a modified genetic algorithm proposed by Yamanaka and Ishida (1996). In fact, the genetic algorithm allows a non-linear inversion analysis to be accomplished that does not depend upon an explicit starting model and allows the identification of the parameter's search space where the global minimum of the inversion problem is.

Since an infinite number of structural models leading to the same H/V spectrum exist, additional information is needed to constrain the inversion. Several publications (Scherbaum et al. 2003, Arai and Tokimatsu 2004, Ohrnberger et al. 2004, Parolai et al. 2006, Castellaro and Mulargia 2009) show that the H/V inversion can provide reliable results once the total thickness of the sedimentary cover is constrained. Although the total thickness of the sediments for the Santiago basin is known (see Figure 2.4) there is, however, some degree of uncertainty in the topography of the bedrock due to the interpolation of the gravimetric data between the measurement sites, since the sites were spaced with an interstation distance of around 500 m. Therefore, the total thickness of the sediments was allowed to vary within 10 % compared to the estimated value derived by the gravimetric data, allowing to avoid problems of trade-off between total thickness and the S-wave velocity profile and to account for uncertainties.

For the inversion, the velocity ranges of the layers were fixed to wide ranges by considering the value intervals for the different sediments in the Santiago basin given by Bravo (1992), Lagos (2003) and Ampuero and Van Sint Jan (2004). In addition, detailed stratigraphic information was used which had been derived from about 150 boreholes distributed over the entire investigated area with depths ranging between 20 m and 300 m (predominantly around 100 m) drilled by the Chilean water company Aguas Andinas S. A. for the determination of the groundwater level. It has been shown (e.g. Arai and Tokimatsu 2004) that sediment thickness and S-wave velocity have the most significant effects on the H/V spectrum, whereas the influence of other parameters such as density (for inversion, density layer parameters were taken from Valenzuela (1978) and fixed) and P-wave velocity is much smaller. Hence, for calculating the P-wave velocity, the relationship of Kitsunezaki et al. (1990) was used:

$$v_p \text{ [m/s]} = 1.11 \ v_s \text{ [m/s]} + 1290. \tag{2}$$

The initial starting population (50 individuals) is generated by a uniform random distribution within the entire parameter space. Throughout this study, the number of layers was prescribed (generally seven; four, if the sedimentary cover thickness is below 50 m). As needed for the computation of the H/V spectral ratios, a fixed standard structural model for the bedrock layer with a high velocity up to 2500 m/s, according to Ortigosa and Lástrico (1971) and confirmed by Bravo (1992), and a density of 2.6 t/m³ (Araneda et al. 2000) was used. Since this inversion is a kind of probabilistic approach using random numbers for finding models near the global optimum solution, the inversions are repeated several times, varying the initial random number.

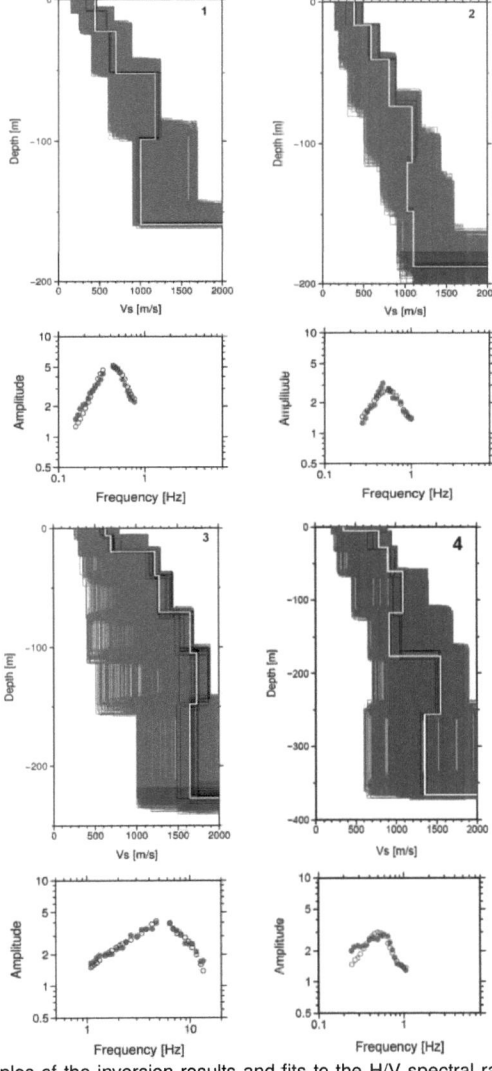

Figure 5.1: Examples of the inversion results and fits to the H/V spectral ratio curves. Top: All four figures show all tested models (grey lines), the minimum misfit model (white line), and all models lying within the minimum misfit + 10 % range (black lines). Bottom: Observed H/V ratio (red dots) and the H/V ratio for the minimum misfit model (open black circles). Numbers indicate sites where the measurements have been carried out and are shown in Figure 3.1.

An example is shown in Figure 5.1, which presents the results of the inversion analysis and the fit of the calculated to observed H/V ratio curves. The presented S-wave velocity profiles provide a rough overview of different velocity structures that are found in the investigated area. The calculated H/V curves fit the observed values well. The reliability of the inversion results is influenced not only by the quality of the input data, but also depends on the frequency range of the H/V curve. However, deeper parts of the model are more influenced by the trade-off between parameters (Scherbaum et al. 2003) and are therefore characterized by a higher uncertainty. This is also confirmed by larger instabilities in the S-wave velocity profiles characterized by misfits within + 10 % of the best model with depth. Although the resolution for deeper parts of the model remains lower, it is still possible to retrieve an average S-wave velocity which might be sufficient for engineering seismology purposes. Additionally, the fact that for more than 85 % of all profiles the total depth of the sedimentary cover differs by only a few percent from the value provided by the gravimetric data set (i.e. the inversion does not provide a value at the margin of the parameter range of the sedimentary cover thickness), indicates the reliability of the underlying gravimetric data.

For validation of the inversion results, we compared the calculated S-wave velocity profiles with other studies (Bravo 1992, GeoE-Tech 2007) in which local velocity information for a few single sites derived from refraction measurements down to a maximum of 30 m is described. Although details about the velocity structure in the uppermost parts of the Santiago basin cannot be retrieved by our inversion technique as the resolution is not high enough for these layers and, therefore, only information about the average S-wave velocity can be retrieved, the average S-wave velocities of these studies are consistent with our data (deviations between values are at most 20 %, often below 10 %).

Although 131 of 146 measurement sites have been used to map the fundamental resonance frequency in Figure 4.16, it is almost impossible to retrieve a reliable S-wave velocity profile for an almost flat H/V curve. So in the end only 125 measurements were found to be suitable for performing an inversion procedure. Only these profiles were used for further analysis.

5.3. Interpolation of the Santiago basin S-wave velocity model

Each of the 125 local S-wave velocity profiles derived from the H/V inversion was then re-sampled with a spatial resolution of one meter, and a 3D S-wave velocity model, parameterized by a rectangular grid with a spatial horizontal resolution of 100 m, was derived. The model contains a detailed description of the sedimentary basin shape derived from gravimetric data as defined by the contact between the lower S-wave velocity sediments and higher S-wave velocity bedrock. Orography, i.e. the shape of the surface, was incorporated, based on digital elevation model data which are available with a resolution of 1 arcsec (approximately 30 m) for all of Santiago and were provided by the Instituto Geográfico Militar.

Different methods have been proposed for the interpolation of spatially changing natural properties (e.g. Isaaks and Srivastava 1989, Wackernagel 1998). A popular method based on an irregularly distributed spatial data set was introduced by Krige (1951). This method uses the local estimation of a parameter based on the weighted spatial mean of samples in the neighbourhood. The weights are defined not only by the distance, but also by the spatial distribution of the samples. Therefore, the accuracy of the estimation depends both on the number of samples and their spatial distribution. A major advantage of the kriging method is the prevention of nugget effects that might occur when dealing with unevenly distributed data. We used a simple ordinary kriging (i.e. the estimation of means in a moving neighbourhood, see Wackernagel (1998) for details) technique which often shows a performance as good as, or even better than, more complex co-kriging methods (Goovaerts 2000) and spline approximations, which are only special cases of the general kriging problem (Laslett 1994).

The performance of the interpolation technique was tested using cross validation (Isaaks and Srivastava 1989). The idea consists of removing one S-wave velocity profile observation from the input data and to re-estimate its value from the remaining data with the help of the described algorithm. The comparison criterion is the mean square error between the former S-wave velocity profile and its re-calculated estimate. The value should be close to zero if the interpolation algorithm is accurate. The procedure is repeated until every sample has been, in turn, removed. The cross validation of this particular form is therefore suitable for assessing the interpolation scheme, because the reliability of the interpolation in under-sampled areas can be verified. Of course, larger variations appear when removing one profile close to the edge of the investigated area, because the lower number of surrounding profiles cannot compensate for the missing data. In general, however, when removing one profile, the mean error between the interpolated S-wave velocity and the values estimated by the inversion algorithm is usually below 10 %; only for the south-western area differences are higher (up to 22 %). It has been found that the S-wave velocity model is stable both for S-wave velocity values close to the surface and for deeper parts of the basin. Due to large discrepancies at the edge of the area of investigation, no interpolation of the S-wave velocity was carried out outside a polygon which is bounded by the outermost measurement sites. By reason of the spatial distribution of the measurements, the S-wave velocity covers an area within a lateral extent of around 270 km².

5.4. Characteristics and interpretation of the 3D S-wave velocity model

To determine the validity of the interpolated model, first a comparison of cross sections of our model with available geological cross sections from literature (Bravo 1992, Iriarte 2003, Iriarte et al. 2006, Pasten 2007) was carried out, although the link between stratigraphy and S-wave velocity is not expected to be direct due to velocity variations that occur between the different kinds of lithology and also inside the same lithology. In all of these studies, the geological units have only been drawn for the uppermost 100 m to 150 m with large lateral uncertainties, especially for deeper parts. As an example, Figure 5.2 shows a comparison of two geological and the corresponding interpolated S-wave velocity cross sections. It has to be mentioned that the geological profiles, presented in different publications, are based on different data sets. Therefore, and due to the large uncertainties, differences in the geologic units of Figure 2.3 and Figure 5.2 can be explained.

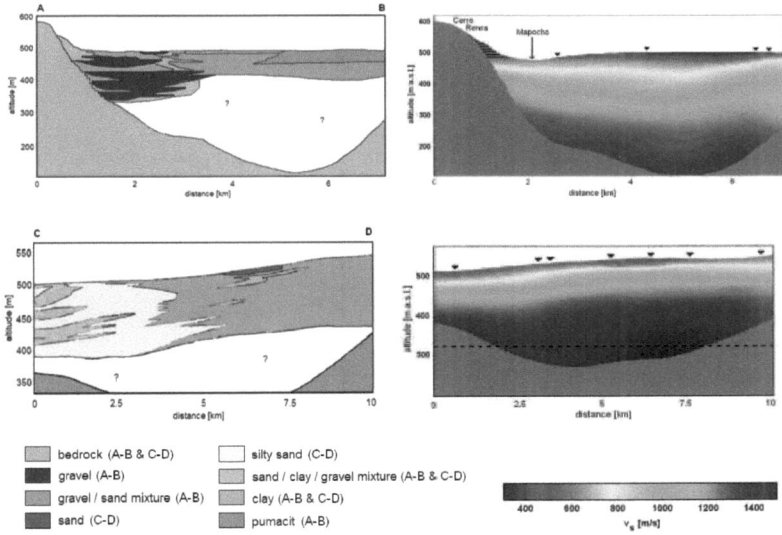

Figure 5.2: Left: Geological cross sections along profiles A-B (upper) and C-D (lower) marked in Figure 2.3 after Iriarte (2003). Right: The corresponding 2D S-wave velocity models calculated by means of interpolating between single S-wave velocity profiles. Black triangles mark the location of the measurement sites within a distance of 500 m to the cross section. For the striped part in the velocity cross section A-B, the results are only due to interpolation. The dashed line in velocity cross section C-D marks the lower border of the geological cross section.

In the southern (i.e. right) part of cross section A-B, the layer of pumacit, characterized by a rather low S-wave velocity, is clearly visible. Our model also shows higher velocities just below the surface for the northern (i.e. left) part of this cross section, an area mainly composed of gravel. Differences in orography between both profiles are due to the different data sets; for the interpolated S-wave velocity model we used data with a high resolution of one arcsec, enabling us to illustrate topographic details like the riverbed of the Mapocho which is cut into the sediments, clearly visible to the left of cross section A-B. For the submontane part of Cerro Renca, the calculated S-wave velocity is only based on interpolation because no measurements have been carried out in the vicinity. Hence, similar to the fundamental resonance frequency (chapter 4.6), the results in the immediate vicinity cannot be considered to be reliable and are therefore masked in Figure 5.2. In profile C-D, the higher velocities of the gravel-sand body, as well as the lower velocities of the sand layer can easily be identified. Therefore, although our model cannot resolve geological details, a general agreement between both is found, allowing us to reliably point out local characteristics that are shown or only foreshadowed in other geological data sets.

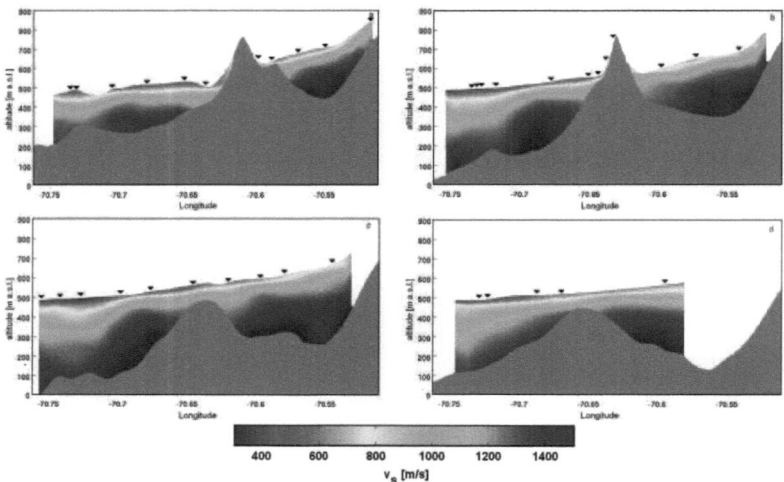

Figure 5.3: West-east cross sections of the interpolated 3D S-wave velocity model within the area of investigation. Black triangles mark measurement sites within a distance of 500 m to the cross section. a) Lat 33.405° S, b) Lat 33.425° S, c) Lat 33.440° S, d) Lat 33.465° S. For all profiles, topography is exaggerated about 16 times. The shape of the bedrock, shown in gray pattern, is derived from gravimetric data. Orography is based on high resolution digital elevation data.

Additional cross sections of our velocity model of the Santiago basin are shown in Figure 5.3. As expected, all profiles show a trend of increasing S-wave velocities with depth, but differences in the velocity gradient can clearly be seen, even at a first glance. However, strong lateral variations can be justified. Figures 5.3 a and b show that to the east of Cerro San Cristobal, the S-wave velocity is higher than to the west. As already mentioned in chapter 4.6, flat H/V curves and H/V spectral ratios of low amplitude are mainly located in the eastern parts of the basin over dense sediments, i.e. Santiago and Mapocho gravel. Up to now it has been assumed (e.g. Bonnefoy-Claudet et al. 2009) that these parts of the basin are characterized by the high seismic velocities of the sediments leading to small velocity contrasts with the bedrock. Our results strongly support this assumption, as well as the distinction between Santiago gravel and Mapocho gravel, as proposed by Valenzuela (1978) and in contradiction to Baize et al. (2006). Mapocho gravel, located in the northern part of the investigated area (see Figure 2.3), is characterised by higher compactness resulting in higher velocities. Also in our model the S-wave velocity for the northern part of the gravel body is increasing slightly faster (Figure 5.3 a around 70.55° W) than for the southern one (Figure 5.3 c between 70.55° W and 70.65° W).

The sub-surface extension of the gravel body is only known approximately and only for the uppermost part with the help of a few boreholes. It has been assumed that the entire gravel body is prograding to the north-west below the fine series, which is also supported by our model. Figure 5.3 b shows relatively high S-wave velocities for the lowermost 200 m of the sediments between 70.67° W and 70.70° W.

To the west of the investigated area (Pudahuel district, see Figure 2.2), the thick layer of pyroclastic flow deposits, outcropping at the top of the filling and characterized by an S-wave velocity being smaller than the velocity of the surrounding materials, can clearly be seen (Figure 5.3 c). Furthermore, the assumption made by Baize et al. (2006) that this low-velocity layer continues at depth towards the south as it may have resulted from a single eruptive phase of the Maipo volcanic complex can be verified. Also in our basin model south of 33.45° S (Figure 5.3 d between 70.75° W and 70.70° W) the S-wave velocity is lower compared to the northern part. The shape of this sequence was previously not imaged in detail.

In general, the model is able to reliably point out local characteristics shown in other data sets, but takes advantage of the fact, that all individual S-wave velocity profiles reach the bedrock, albeit the resolution for deeper parts of the model remains lower. Furthermore, all recent publications (Baize et al. 2006, Pasten 2007, Bonnefoy-Claudet et al. 2009) distinguish only three soil categories: gravel, fine deposits, and ashes for the Santiago basin, claiming that only these categories behave differently from each other under seismic excitation. However, it has been shown that this simplification does not hold, as strong lateral S-wave velocity differences, even within the single units, occur and therefore also no simple depth – S-wave velocity relation, as suggested by Bravo (1992) and Pasten (2007), may exist. Furthermore, the complexity of the lithological succession, characterized by sharp lateral and vertical variations of the physical properties and of the S-wave velocity gradient over short scale, may even hint, that not only depositional processes, but also tectonic activity account for this complexity (Magdalena Scheck-Wenderoth, personal communication).

5.5. Correlation between slope of topography and v_s^{30}

As our model can provide reliable information about the S-wave velocity structure of the basin, and topographic data for the entire basin are available at a high resolution, we are able to check if a simple and inexpensive technique suggested by Wald and Allen (2007), based on a supposed correlation between the slope of topography and the superficial S-wave velocity, exists on a local scale and might be useful for soil classification in Santiago de Chile.

First, the slope of the topography was calculated for the entire area under investigation using the highest possible resolution of 1 arcsec. As the elevation in the Santiago basin steadily increases slightly from west to east, with only a marginally rising gradient, most of the investigated area is characterized by a gentle slope gradient less than 0.03 (i.e. vertical rise over horizontal distance traversed, Figure 5.4). Extended hills like Cerro Renca and Cerro San Cristobal outcropping from the sedimentary filling and characterized by steep slopes can easily be seen, while even the riverbed of the Mapocho can be identified.

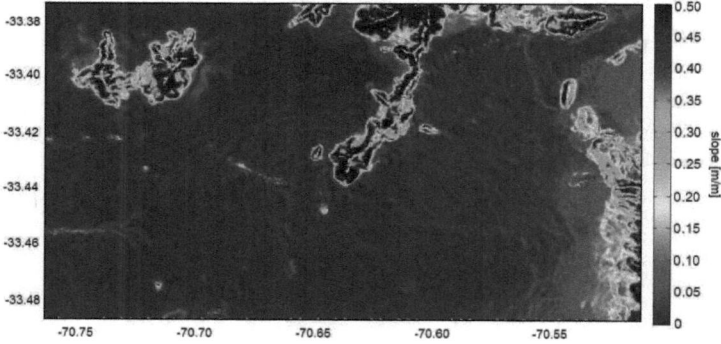

Figure 5.4: Topographic gradient of the area under investigation, calculated using a topographic resolution of 1 arcsec.

To compare the topographic gradient with v_s^{30} values, the relationship is characterized in terms of discrete steps in S-wave velocity according to the U. S. National Earthquake Hazards Reduction Program (NEHRP) (Building Seismic Safety Council 2004). To increase resolution, the NEHRP boundaries are further subdivided into narrower velocity ranges, following Wald and Allen (2007).

As it is the aim of the study to verify if any relationship between slope and v_s^{30} exists using data collected on a local scale, we first correlate v_s^{30} with topographic slope (resolution 1 arcsec) at each v_s^{30} measurement site. Subsequently, topographic slope at any site that falls within the subdivided NEHRP ranges is assigned a v_s^{30} that defines the median value of the grouped NEHRP windows, following Wald and Allen (2007). Afterwards, the slope ranges are subdivided allowing the best fit to the median values of the subdivided NEHRP windows. Results are listed in Table 5.1 and shown in Figure 5.5 a. Although the limited number of only 125 measurement sites will allow only a rough estimation of the correlation coefficients, tendencies should be verifiable. While large parts of the Santiago basin are characterized by slope values not differing too much from each other (Figure 5.4), the variability in the resulting v_s^{30} is high. Hence, although Figure 5.5 a shows the best fit of variable slope ranges to the given v_s^{30} categories, no obvious correlation between high-resolution slope information and v_s^{30} can be seen, and many data points fall outside the calculated ranges.

Table 5.1: v_s^{30} and associated topographic slope ranges calculated according to Wald and Allen (2007) from measurements carried out in Santiago de Chile for a topographic resolution of 1 arcsec and taken from Allen and Wald (2009) for active tectonic regions using a 9 arcsec resolution data set. Only NEHRP v_s^{30} ranges found in the Santiago basin are listed here.

NEHRP site class	v_s^{30} range [m/s]	Slope range calculated from measurement data (Figure 5.5 a)	Slope range given by Allen and Wald (2009) (Figure 5.5 b)
DE	180-240	no values	0.00030-0.00035
D	240-300	no values	0.00035-0.010
DC	300-360	<0.0085	0.010-0.024
CD	360-490	0.0085-0.013	0.024-0.08
C	490-620	0.013-0.033	0.08-0.14
CB	620-760	0.033-0.096	0.14-0.20
B	>760	> 0.096	>0.20

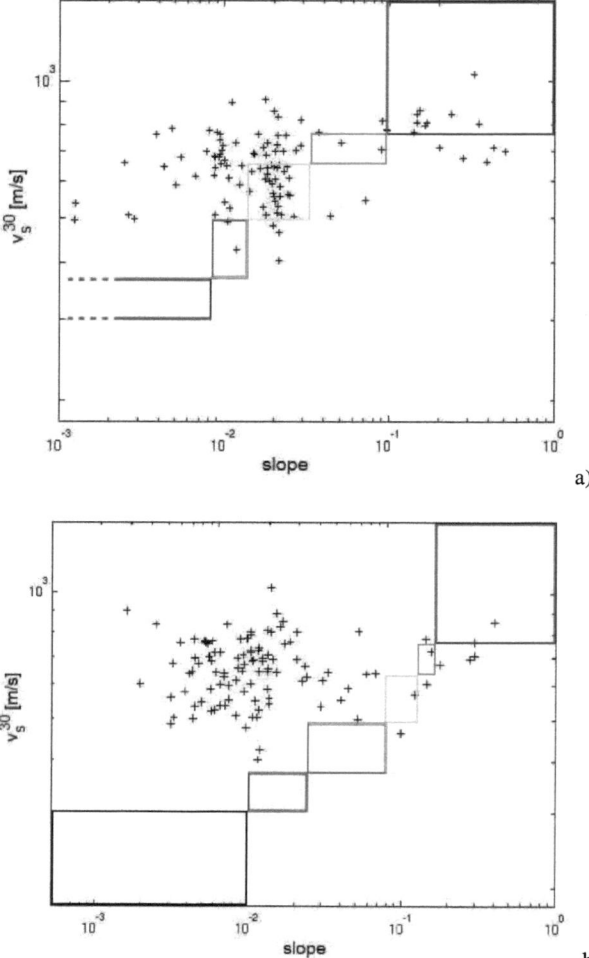

Figure 5.5: Correlation of measured v_s^{30} against slope of topography for different topographic resolution. Color-coded polygons represent v_s^{30} values and their associated slope ranges listed in Table 5.1. Top: Topographic resolution of 1 arcsec and the re-calculated slope ranges for given v_s^{30} categories. Bottom: Topographic resolution of 9 arcsec. Values of associated slope ranges are taken from Allen and Wald (2009) for active tectonic regions. See text for further discussion.

Of course, the application of lower resolution topographic data will result in different slope values. Allen and Wald (2009) base their results on a topographic resolution of 9 arcsec; the ranges are also listed in Table 5.1. When using the suggested slope-v_s^{30}

ranges for the Santiago data set with a topographic resolution of 9 arcsec the agreement is even smaller (Figure 5.5 b). However, this extreme divergence cannot completely be explained by the fact of using different topographic resolutions. In fact, when using a lower resolution of only 9 arcsec or even 30 arcsec for calculating the topographic gradient at each measurement site, as has been done by Allen and Wald (2009), steep slopes over short distances will not be resolved due to the lower sampling rates inherently smoothing maximum slope values inferred from the topography. Note, however, that the measured v_s^{30} values in Figure 5.5 b are still much higher than the expected ranges of Allen and Wald. Only for steeper slope ranges an agreement with the given velocity range is found. The expected trend that materials characterized by high velocities are more likely to maintain steeper slopes and vice versa can only benevolently be seen.

On the other hand, when using a lower resolution, high slope values (approximately greater than 0.1) will tend to be scaled down, which can also be seen when comparing Figures 5.5 a and b, leading to even less agreement. Hence, independent of the spatial resolution used, our results reduce any confidence in the applied correlations between topographic gradient and v_s^{30}. Even lower resolution data (30 arcsec) does not provide comparable estimates of v_s^{30}, although the relief in the Santiago basin is quite low.

However, studies dealing with estimating local site conditions mainly focus on densely populated areas with a moderate extension. Therefore, there is the utmost necessity to find a technique giving a reliable overview over regional scales which can further be applied with confidence on a local scale. At least to reduce the overall bias for Santiago de Chile, a general shift to lower slope ranges for the associated v_s^{30} values would therefore be necessary. But in contrast, Wald and Allen (2007) observed exactly the opposite trend for the Salt Lake City region. Therefore, it seems that a general rule cannot be found.

5.6. Correlation between S-wave velocity and macroseismic intensity of the 1985 Valparaiso event

The reason for the apparent lack of coincidence in any correlation between v_s^{30} and the topographic slope might be clarified, when analysing the geological situation: A large part of the Santiago basin is characterized by a low, but measurable topographic gradient with only a limited number of outcropping bedrocks characterized by steep slopes (Figures 5.4). In contrast, the surface geology is rather heterogeneous (Figure 2.3), indicating that the deposition played a minor role in differentiating the material and their friction angles. Therefore, one might check, if a first order correlation between v_s^{30} and surface geology may be detectable. To this regard, we mapped v_s^{30} for the area under investigation using a 2D kriging algorithm similar to the one described in chapter 5.3. Again, we calculated v_s^{30} only for a polygon which is bounded by the outermost measurement sites. Results are shown in Figure 5.6.

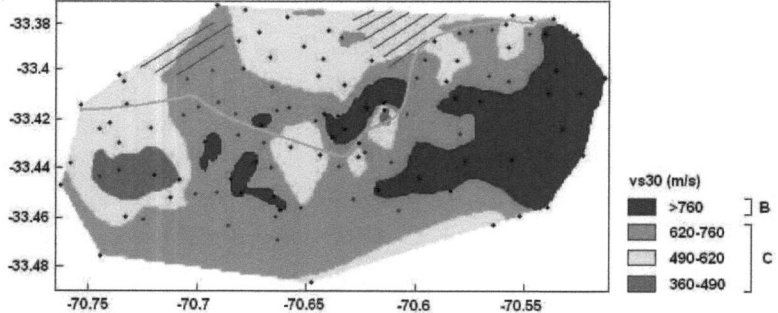

Figure 5.6: v_s^{30} (NEHRP classification) for the area limited by a linear connection between the outermost measurement sites. The turquoise line represents the Mapocho river. Black dots show the location of the ambient noise measurement sites used for mapping v_s^{30}. In the dashed area, no measurements were carried out. Therefore, the results are only due to interpolation and are not plausible as outcropping bedrock can be found there.

Areas characterized by different S-wave velocities in the uppermost 30 m can clearly be identified. In the western part an ellipsoidal area of low v_s^{30} values represents the layer of pumacit on the top of sedimentary filling (see Figure 2.3). In the central part, the outcropping igneous bedrock of the Cerro San Cristóbal is apparent. The v_s^{30} values found there contrast with those in the surroundings and depict the shape of the hill quite well. Note, that in the northern part of the Cerro San Cristóbal and also close to the Cerro Renca, no ambient noise measurements have been carried out. Therefore, no reliable v_s^{30} interpolation values are found for these parts of the investigated area and are thus dashed in Figure 5.6.

Some individual measurement sites in the central and the north-eastern part located on gravel show low v_s^{30} values where higher velocities would have been expected. These differences confirm that within the same lithology strong lateral variations of S-wave velocity can be found. The changes are explainable when bearing in mind the course of the Mapocho river. In accordance with Pasten (2007), the river beds in the Santiago basin and the associated flow velocity are mainly controlled by topography. As a result, where the flow velocity is low, sediment drops out of the flow and deposits. Therefore, this emphasises that geology-based values should not be taken as constants within a specific geological unit as changing material properties (e.g. particle size) often appear.

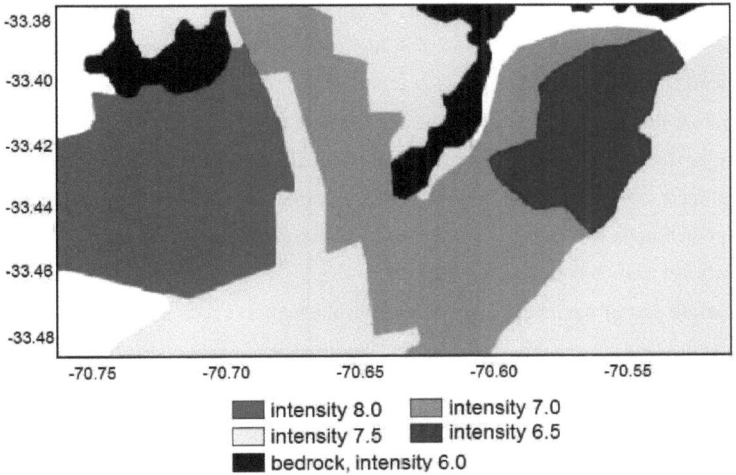

Figure 5.7: Intensities (MSK scale) for the 1985 earthquake of the investigated area. Values are taken from Astroza and Monge (1991). In the white parts no reliable intensity values could be determined due to the limited amount of information available.

When comparing our v_s^{30} map with the MKS intensity values determined for the Mw=7.8 (epicenter 120 km west of Santiago) earthquake of 3 March 1985 (Astroza and Monge (1991), Figure 5.7), a tendency can be noticed that the intensity distribution for that event was also influenced by local variations of v_s. Moreover, although the exact intensity distribution for the 2010 Maule event has not been mapped in detail yet, we found that most of the damages for this event were concentrated in the north-western part of the city where a small-sized low-quality building stock can be found (Astroza et al. 1993). At first glance and also outlined in chapter 4.6, this cannot be explained by an accordance of the fundamental resonance frequencies of the soil and the building stock, but one should keep in mind that higher harmonics also might correspond to the eigenfrequencies.

Nevertheless, the fundamental resonance frequency map (Figure 4.16) can provide complementary information, since it shows a slight tendency that large H/V peak values are observed in areas with higher intensities (western part of the city) and small H/V peak values in areas with lower intensities (eastern part). Measurements carried out on Cerro San Cristóbal, showing high fundamental frequencies and high H/V peak amplitudes, do not follow this trend. Nonetheless, a more detailed analysis is not possible. On top of the pumacit layer, where the intensity reached a maximum, the lowest v_s^{30} values (Figure 5.6) and also high impedance contrast values between the sediments and the bedrock (Figure 5.3 c and d) are found but, in contrast, only some H/V ratios show high amplitudes. The highest H/V peak amplitudes are found for the central northern part of the investigated area (Figure 4.16). Summarizing, considering the outlined results, no direct but only a slight correlation between H/V amplitude and damage is found. This result is not surprising, since damages depend on ground motion intensity but also on building fragility.

The eastern part of the investigated area shows the highest v_s^{30} and low intensity values. Of course, only a general trend can be identified when mapping v_s^{30} and no strict borders should be drawn between individual units. Different standard spectral ratio amplitudes at sites of our network showing the same v_s^{30} range were found as well as same SSR amplitudes for different categories (Table 5.2), indicating the incompleteness of v_s^{30} alone to fully describe site effects and the need to consider additional parameters.

Table 5.2: v_s^{30}, associated NEHRP category, and SSR amplitude (S-wave, see Figure 4.10) for sites of the 2008 network.

Station number	v_s^{30} [m/s]	NEHRP site class	SSR amplitude
S1	1040	B	reference station
S2	454	CD	2.6
S3	580	C	< 2
S4	703	CB	< 2
S5	645	CB	2.2
S6	630	CB	7.8
S7	709	CB	2.3
S8	792	B	2.2

When comparing Figures 2.3, 5.4, and 5.7, it seems obvious that, except for the outcropping of igneous bedrock, no clear correlation between evaluated intensities and topographic slope can be found for the 1985 event. The surface geology offers a very poor first-order indication of intensity distribution, whereas v_s^{30} might allow a better, although only approximate, distinction of areas characterized by different intensity values. Of course, v_s^{30} is not the only factor controlling ground-motion amplification. In addition to near surface site conditions seismic waves are also known to be significantly influenced by topographic effects and the basin structure. The knowledge about the basin outlined so far allows a detailed modelling of seismic wave propagation and calculation of seismic hazard scenarios.

Chapter 6

Simulation of the Santiago basin response by numerical modeling of seismic wave propagation

6.1. Introduction

In recent years, 3D simulations of earthquake ground motion using more realistic velocity structures have seen a rapid advancement. The numerical methods used range from finite difference (e.g. Olsen et al. 1995, Graves and Wald 2004, Hartzell et al. 2006, Harmsen et al. 2008) to finite element (e.g. Bao et al. 1989, Aagaard et al. 2001, Bielak et al. 2005) and to spectral element techniques (e.g. Faccioli et al. 1997, Komatitsch et al. 1999, Paolucci et al. 1999, Tromp et al. 2005, Stupazzini et al. 2009).

In general, all these techniques are able to simulate earthquake ground motion, although no single method can be considered as the best one for all medium-wave field problems regarding computational efficiency and numerical accuracy (Chaljub et al. 2010). Especially in the near-fault range, the simulation of earthquake ground motion depends on several parameters like the focal mechanism, rupture propagation characteristics, and the complex interaction with the shallow structure. Therefore, considering the mentioned effects from a numerical point of view implies large-scale 3D simulations of seismic wave propagation paying particular attention to the surface topography and sedimentary basin parameters. Up to now, such basin structures and their influence on seismic wave propagation are largely illustrated through numerical studies (e.g. Horike et al. 1990, Graves 1993, Hisada et al. 1993, Moczo et al. 1999, Frankel et al. 2001, Hartzell et al. 2006, Olsen et al. 2006) and observations (e.g. Kagawa et al. 1992, Kinoshita et al. 1992, Phillips et al. 1993, Frankel 1994, Hatayama et al. 1995, Field 1996).

To accommodate the considerable surface topography in the Santiago metropolitan area as well as the variable low wave speed sedimentary cover in the basin, numerical simulations are carried out making use of the spectral element method (SEM) software package GeoELastodynamics by Spectral Elements (GeoELSE), jointly developed by the Center for Advanced Research, Studies, and Development in Sardinia (CRS4) and by the Department of Structural Engineering of the Politecnico di Milano, Italy (Stupazzini 2004). In a very general view, the SEM can be seen as a generalization of the finite element technique, based on the use of high order piecewise orthogonal polynomial functions. The method was first used in the 1990s to model seismic wave propagation in local and regional surroundings (e.g. Priolo et al. 1994, Faccioli et al. 1997, Komatitsch and Tromp 1999); more recently it was also extended to global wave propagation (Komatitsch and Tromp 2002a, 2002b, Capdeville et al. 2003, Chaljub et al. 2003).

The key features of the SEM discretization are as follows:

1. Like in the finite element method (FEM), the entire model volume Ω is subdivided into a number of non-overlapping elements, such that $\Omega = \bigcup_{k=1}^{N} \Omega_k$.
2. The spatial integration is performed based upon the Gauss-Lobatto-Legendre quadrature, while most classical FE techniques use Gauss quadrature.
3. The expansion of any function within the elements is accomplished based upon Lagrange polynomials of suitable degree n constructed from $n+1$ interpolation nodes.
4. In each element, the interpolation nodes are chosen to be the Gauss-Lobatto-Legendre points, i.e. the $n+1$ roots of the first derivatives of the Legendre polynomials of degree n.

Specifically, the combined use of Lagrange orthogonal polynomials and of the Legendre-Gauss-Lobatto quadrature rule represents one key aspect of the implementation of the spectral element approach in GeoELSE. So the main advantage of the SEM is that it combines the flexibility of the finite element method with the accuracy of pseudospectral techniques. Moreover, the method is also capable of providing an increase of accuracy by simply increasing the spectral degree of the functions, i.e. their algebraic degree. So this procedure is completely transparent to the user. On the other hand, like for the standard finite element approach, also grid refinement can further be used to improve the accuracy of the numerical simulations. Altogether, the SEM can be seen as part of the so-called h-p method (Faccioli et al. 1996) where h refers to the grid size and p to the algebraic degree. The exponential accuracy of the method is ensured and the computational effort is minimized because the resulting mass matrix is exactly diagonal. This approach requires, however, that the mesh elements are hexahedral resulting in much effort to be spent on an accurate mesh generation.

In particular, the meshing strategy has to account for the true positions of the material interfaces and the fault. The accurate representation of the real geometry is generally assumed to be important to achieve high-quality simulation results avoiding numerical artefacts due to inappropriate model discretization. However, for the Santiago basin, showing laterally variable sedimentary layers and sharp transitions between shallow sediments and the underlying bedrock, the generation of detailed meshes poses a formidable challenge. This task was successfully solved by means of the software CUBIT (http://cubit.sandia.gov) which incorporates a set of advanced meshing schemes specifically developed to handle the hexahedral unstructured meshing problem.

To verify that the underlying models are correct and might be used for further analyses, an excellent opportunity is offered by the 27 February 2010 Maule event whose aftershocks have been recorded by a dense temporary seismic network installed in the city (see Figure 3.1 and Table 3.2). In general, a sufficiently high level of agreement or a sufficiently small level of misfit between recorded data and theoretical prediction can be considered as a confirmation of a theoretical model of an investigated process. In particular, the agreement between recorded and numerically predicted earthquake motion can be considered as an ultimate criterion for the capability to simulate earthquake ground motion.

Therefore, it will first be shown that the characteristics of the model of the Santiago basin are able to account for different site conditions. The question whether numerical modeling of wave propagation can explain peculiarities of wave propagation and amplification found in the analyses of empirical data will be addressed. Furthermore, as has already been outlined in chapter 2.2, nearby potential sources like the San Ramón Fault constitute a latent seismic hazard for the city. To this regard, although the numerical prediction of ground motion cannot yet be considered mature, it will be analyzed how the structure of and the topography in the Santiago basin as well as the hypocenter location affect near-fault ground motion.

6.2. The spectral element numerical code GeoELSE

The 2D / 3D numerical code GeoELSE is based on the SEM formulation proposed by Faccioli et al. (1997) and is specifically designed to effectively perform linear and non-linear visco-elastic seismic wave-propagation analyses, including the combined effect of a seismic fault rupture, the propagation path, and a complex geological feature in the simulation.

As in the finite element approach, the dynamic equilibrium problem is stated in the weak form through the principle of virtual work (Zienckiewicz and Taylor 1989) and, through a suitable discretization procedure, is written as an ordinary differential equation system with respect to time (Chaljub et al. 2007).

Absorbing boundaries are implemented by means of the conditions proposed by Stacey (1988); the free-surface condition is a natural condition in the SEM. The visco-elastic behavior implemented in GeoELSE is based on replacing the initial term $\rho \frac{\partial^2 u}{\partial t^2}$ in the wave equation by $\rho(\frac{\partial^2 u}{\partial t^2} + 2\gamma \frac{\partial u}{\partial t} + \gamma^2 u)$ with u being the generic displacement component, ρ the density, and γ the attenuation parameter. Kosloff and Kosloff (1986) have shown that, with such a replacement, all frequency components are attenuated equally, resulting in a quality factor that is frequency proportional $Q = Q_0 \frac{f}{f_0}$, where $Q_0 = \pi f_0 / \gamma$ and f_0 is a average frequency value representative of the frequency range to be propagated. In particular, when f is near f_0, the approximation to a constant Q works very well. Following Graves (1996), the reference frequency f_0 is chosen near the peak frequency of the source and the simulation bandwidth is centered around this value.

To account for the non-linear behavior of the soil, a non-linear visco-elastic soil model was implemented in GeoELSE by means of a generalization of the 1D load conditions of the classical G-ε and D-ε curves (Kramer 1996) to three dimensions. G, D, and ε represent the shear modulus, the damping ratio, and the 1D shear strain. To this regard, a scalar measure of the shear strain amplitude is calculated at the generic position and the generic instant of time. Afterwards, the shear strain value is introduced in the G-ε and D-ε curves, and, subsequently, the corresponding parameters are updated for the following time step. Therefore, unlike in the classical equivalent-linear approaches, the initial values of the dynamic soil properties are recovered at the end of the excitation. Since no iteration is required at the generic time step, this non-linear implementation implies only a moderate increase of computer time of about 20 % compared to the linear case (Stupazzini et al. 2009).

Several studies have shown that ground motion complexity arises mostly from the details of the rupture process (e.g. Beresnev and Atkinson 2002, Mai et al. 2006), especially in the near-field of the seismic source, within a maximum distance varying from 1-2 km to few tens of kilometers, depending on the magnitude of the earthquake. Therefore, synthesizing reliable ground motion time histories for both engineering and seismological purposes cannot disregard a proper characterization and modeling of the fault rupture process. Although significant effort has been devoted to the study and application of dynamic modeling of extended sources (Guatteri et al. 2003, Rippenger et al. 2007), dynamic fault models are still at research stage due to their complexity and computational costs. The computational limits can be overcome to some degree; however, these models have been restricted to lower frequencies (less than 2-3 Hz). To this regard, kinematic modeling, like represented in GeoELSE, remains the best trade-off between computational cost and adherence to the physics of fault ruptures (see e.g. Hall et al. 1995, Kamae et al. 1998, Pitarka et al. 2000, Hartzell 2002, Archuleta et al. 2003).

Of course, given the complexity and inevitable uncertainty of the underlying models, GeoELSE had first to be validated with several analytical solutions (e.g. Faccioli et al. 1997, Stupazzini 2004). The seismic response of the valley of Grenoble (French Alps), showing a complex 3D geometry and large velocity contrasts, has recently been the object of an international benchmark indicating a proper implementation of the SEM in GeoELSE (Chaljub et al. 2010).

6.3. Test for accuracy and stability – the 1 April 2010 aftershock

6.3.1. Implementation

6.3.1.1. Mesh geometry

The numerical calculation presented here refers to an aftershock occurring on 1 April 2010 at 12:53:07 UTC. This event (M=5.2) was located around 180 km south-west of the city (latitude 34.65° S, longitude 71.82° W) at a depth of around 11.8 km and recorded by the 2010 temporary seismic network. Since our interest consists in modelling the sedimentary basin behavior the meshing strategy is based on two standard steps: First of all, the entire volume is cut into small slices and then each slice has to be meshed with a standard meshing scheme. As mentioned in chapter 6.1, the choice of the grid size h is related to the algebraic degree p. In spite of complex material interfaces, for resolving the wave field for the shortest period, the number of grid points per wavelength should be at least equal five. As wave speeds increase with depth, the mesh should be denser close to the surface and close to low-speed regions. However, on structured meshes the resolution can be reduced for S-wave speed ratios smaller than 0.5 (Pelties et al. 2010) which also applies for the Santiago basin.

The final mesh is depicted in Figure 6.1 and consists of 1,844,014 elements, the size of which ranges from a minimum of about 60 m (inside the sedimentary basin) up to 900 m. The mesh is designed to propagate frequencies up to 1.3 Hz with p = 3 (51,632,392 nodes) and up to around 1.8 Hz with p = 4 (119,860,910 nodes).

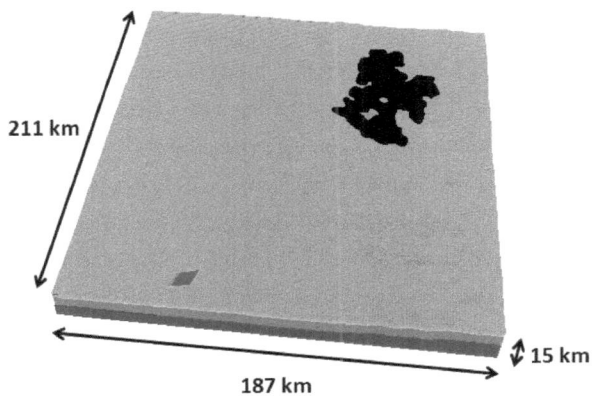

Figure 6.1: 3D hexahedral spectral element mesh adopted for the computation of the 1 April 2010 aftershock with the GeoELSE software. For simplicity, the spectral elements are shown without the Legendre-Gauss-Lobatto nodes. Different colors refer to different mechanical properties (see chapter 6.3.1.2). The shape of the Santiago basin is shown in black. The location of the fault is represented by a gray rectangle.

Non-linear effects were neglected for this analysis, since ground strains induced by the moderate shaking are relatively small due to large source to site distances.

6.3.1.2. Santiago basin model

The surrounding topography of the large-scale numerical model is constructed based on the GTOPO30 data set, a 30 arcsec digital elevation model (available at http://eros.usgs.gov/#/Find_Data/Products_and_Data_Available/gtopo30_info). For the Santiago metropolitan area, digital elevation data are available with a resolution of

30 m. As already mentioned, the shape of the sediment bedrock interface is based on data from gravimetric measurements (Araneda et al. 2000).

The impossibility of finding any simple depth – S-wave velocity relation for our data set has been mentioned in chapter 5.4. However, due to computational limitations we had to assume that the dynamic properties of the sedimentary basin only vary with depth. To this regard, v_s [m/s] = 400 + 55 $z^{1/2}$ with depth z (measured in meters) allows the best fit of the numerical model to the calculated data and was used for the calculation of v_s in the entire basin. For the P-wave velocity we assumed, consistent with equation (2) in chapter 5.2, v_p [m/s] = 1730 + 60 $z^{1/2}$. The depth dependency of the density is taken from Bravo (1992): ρ [kg/m³] = 2100 + 0.15 z. Inside the sedimentary basin, the smooth vertical variation is taken into account, assigning each Legendre-Gauss-Lobatto point the mechanical properties evaluated according to the prescribed depth variation.

The bedrock is layered with v_p = 4700 m/s and v_s = 2400 m/s between 0 and 2.2 km depth, v_p = 5900 m/s and v_s = 3200 m/s between 2.2 km and 8.9 km depth, and v_p = 6200 m/s and v_s = 3450 m/s below 8.9 km depth (Godoy et al. 1999, Barrientos et al. 2004). The frequency f_0 was set to 0.5 Hz in this study. As described in chapter 6.2, the quality factor Q is almost constant around f_0. However, low frequencies smaller than 0.5 Hz will therefore be slightly over-damped, whereas higher frequencies will be enhanced in the sedimentary layer. Stupazzini et al. (2009) obtained good agreement when comparing the waveforms of such a model with a constant Q model, especially in terms of peak values and the overall shape of the waveforms, and concluded that such agreement tends to deteriorate only in the coda part of the signal, where the effect of the different Q assumptions becomes more important.

6.3.1.3. Treatment of the kinematic source

The numerical simulation corresponds to an event with a magnitude of 5.2, geometrically defined by a 4.7 km x 3.5 km rectangle (Wells and Coppersmith 1994). The rupture is in plane with a constant slip density. The parameters of the source mechanism (strike = 195°, dip = 58°, rake = -142°) are taken from CMT catalog (http://www.globalcmt.org). The rupture circularly propagates from the hypocenter located in the center of the rectangle with a rupture velocity of v = 2800 m/s. The time dependency of the seismic moment tensor source is described by an approximate Heaviside function, i.e.

$$M_0(t) = \frac{1}{2}\left[1 + \mathrm{erf}\left(2\frac{t-2\tau}{\tau/2}\right)\right]. \tag{3}$$

where erf (•) is the error function and τ is the rise time, measured as the time necessary to attain 5 to 95 % of the final slip (Mai 2009). The values are selected that the slip velocity is around 1 m/s. The total number of spectral nodes required to model the fault is around 350.

6.3.2. Comparison of numerical predictions

A comparison of velocity waveforms recorded at and simulated for the sites of the receivers of the 2010 seismic network (see Table 3.2) are shown in Figure 6.2. The waveforms have been band-pass filtered (0.05 Hz < f < 1.8 Hz).

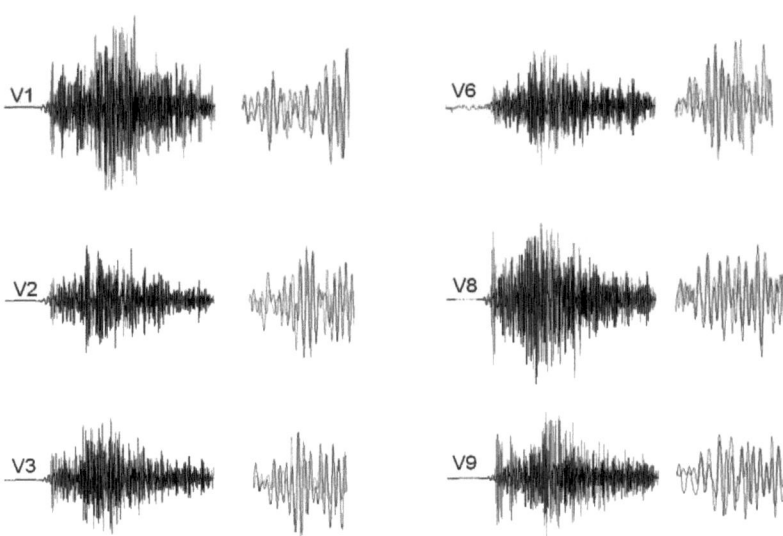

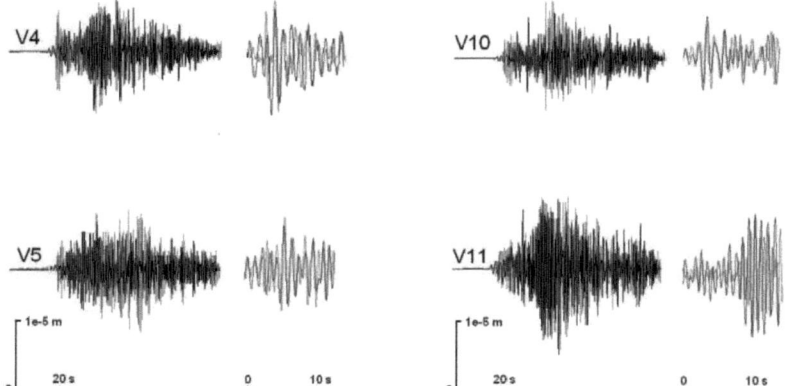

Figure 6.2: Comparison of recorded (red) and simulated (black) vertical component velocity waveforms of the 1 April 2010 earthquake for ten sites of the network (for stations V7 and V12 the earthquake has not been recorded in sufficient quality). Station locations are indicated in Figure 3.1 and listed in Table 3.2. The velocity waveforms are band-pass filtered between 0.05 and 1.8 Hz. For each station, the left figures show the entire waveform and the right figures represent parts of the waveform around the S-wave arrival in close up.

In general, the recorded data can be fit reasonably well on all three components in terms of amplitude and duration. Although we had to assume that the dynamic properties of the sedimentary basin only vary with depth and not laterally, stations in the north-western part of the studied area on the top of low-velocity soils (i.e. stations V1, V8, V9, and V11) in general show higher amplitudes and longer shaking durations. On the contrary, for stations in the central and the eastern part of the study area (V2, V4, V6, and V10) smaller amplitudes and shorter shaking durations are found, similarly to chapter 4.2.1. Although located on hard rock, topographic effects might cause that V3 does not show the smallest amplitude.

Particularly noticeable discrepancies between the simulations and recordings appear for the P-wave arrival. This might result from the use of a frequency proportional quality factor (see chapter 6.2) which causes low frequencies to be overdamped and high frequencies to be enhanced.

However, the time series themselves should be used with caution, since there is no guarantee that the spectrum of each simulation is close to the real spectrum. To accomplish this, acceleration response spectra at frequencies from 0.05 to 1.8 Hz, derived each from the simulation and from the recordings, are compared in Figure 6.3.

In general, the curves are in good agreement both for shorter and longer periods. Curves on thick sediments in the north-western part of the studied area show higher amplitudes at higher frequencies, whereas the amplitudes for sites in the eastern part of the basin are generally lower for small frequencies. Combined with the findings of Figure 6.2 outlined above, this might indicate that the vertical variation of the dynamic properties within the sedimentary basin has been chosen correctly on the whole.

On the contrary, at station V9 the simulation underpredicts the observed spectrum; this trend can also be seen in the seismogram of V9 in Figure 6.2: The energy content of the seismogram in the simulation is significantly smaller than in the recording. The mismatch at this site might be due to a significant difference between the dominant dynamic soil properties and the interpolated ones used in the simulation.

At station V3 the mismatch of the acceleration response spectra for higher frequencies might be explained by pronounced small scale topographic variations in the surroundings of station V3. These variations have not been captured by the mesh but might cause significant amplification at higher frequencies, in consistency with Géli et al. (1988).

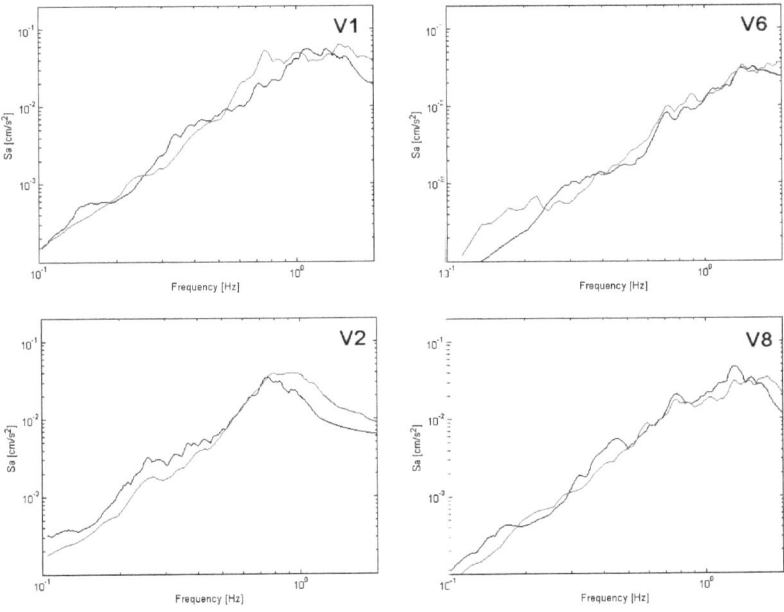

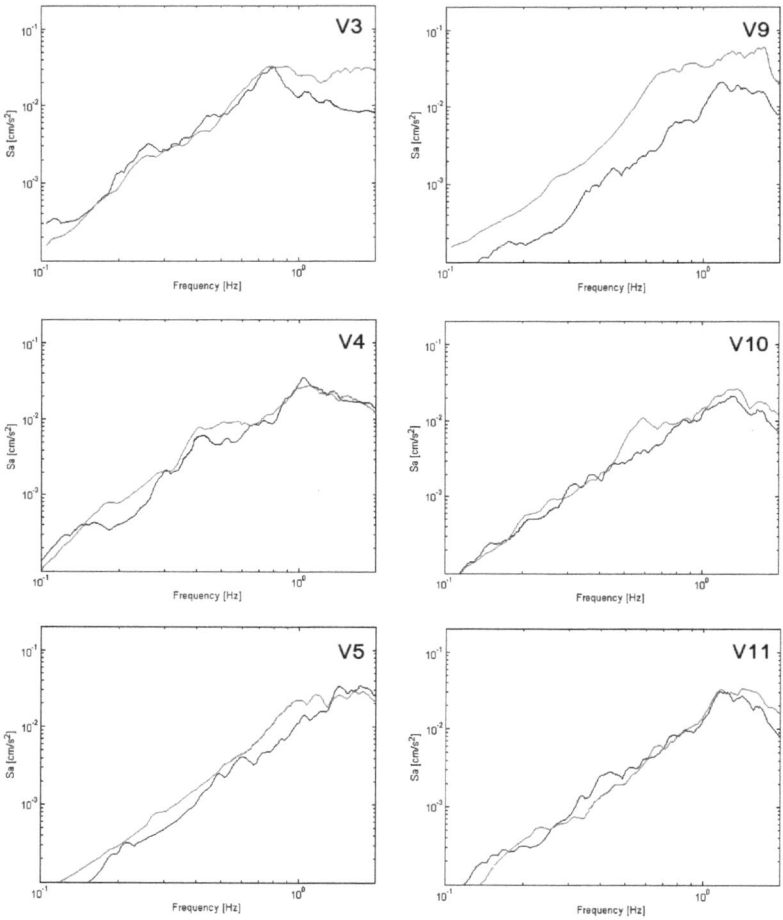

Figure 6.3: Comparison of the observed 5 % damped acceleration response spectra (red) with the spectra simulated in this study (black) for the vertical component of ground motion for the M=5.2 event on 1 April 2010.

Anyhow, although we have included a complex model in the simulations, considering topography and the shape of the sediment-bedrock interface, one cannot hope to match all details of the recorded signals, also because of the lack of detailed information about the rupture process. The generalized rupture model of this earthquake that has been used incorporates only limited information. Altogether, the discrepancies might be due to the combined effects of a properly characterized seismic source and a more detailed shallow crustal model, including near-surface complex geological irregularities which often significantly modify ground motion (e.g. Anderson et al. 1996). An additional difficulty in basin simulations is the lack of detailed knowledge of attenuation. However, since we found a satisfactory agreement in terms of amplitude and especially in terms of duration, we assume that the underlying assumption regarding attenuation is quite realistic. Furthermore, our current basin model is largely based upon only vertically varying S-wave speed information. Therefore, the basin model could further be improved by adding lateral S-wave velocity variations and further constraints on P-wave velocity for the basin and the subregions of the model.

6.3.3. Effect of basin depth and surface topography

It is not only the properties of the low-speed sedimentary basin but also the pronounced surface topography and the shape of the sediment bedrock interface that significantly distort the propagation of waves. Figure 6.4 shows snapshots of the aftershock simulation for the vertical component of the velocity wave field (band-pass filtered between 0.05 Hz and 1.8 Hz). The P-wave reaches the south-western edge of the basin with relative small amplitude at approximately 25 s. The wave fronts slow down and are influenced due to the low shear wave speed in the basin. After around 40 s, the slower but stronger S-wave reaches the edge of the basin and propagates further with significantly varying amplitude.

Especially in the north-western part of the basin the ground motion is very complex. The Cerro Renca (see Figure 2.2), a hill rising more than 200 m above the surrounding plane, causes the seismic waves to be largely reflected and scattered. (For the numerical simulations the topography of the hill has been considered correctly. However, its exact boundary is not visible in Figure 6.4 but indicated with letter A.) To the east of Cerro Renca and to the west of Cerro San Cristóbal strong ground motion occurs which might be due to constructive interference between different arrivals reflected at the facing hills (labelled B in Figure 6.4). As already outlined in chapter 4.6, serious structural damage was found for this part of the city following the 2010 Maule earthquake. The results found here suggest that the trapping of seismic energy might have caused this impact. Also to the south-west of station V1 (green quadrate in Figure 6.4) and between Cerro

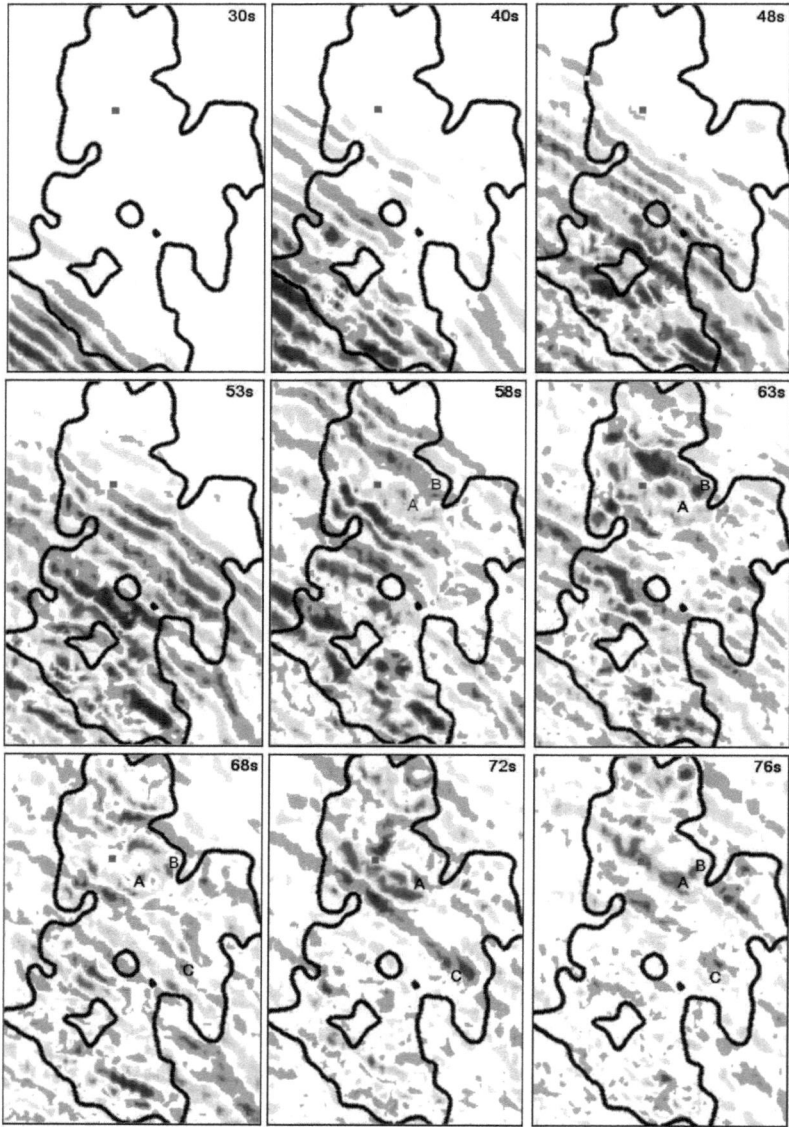

Figure 6.4: Snapshots of the vertical component of the simulation of the 1 April 2010 earthquake. Red colors indicate positive velocities, and blue colors indicate negative velocities. The black lines mark the border between the sedimentary basin and outcropping bedrock. The green quadrate indicates the position of station V1. In the basin, the waveforms are clearly distorted and amplified. See text for further discussion.

Chena and Cerro Lonquén (for the exact location of these hills see Figure 2.2) in the south-west of the basin, interference and amplification effects are clearly visible between 53 s and 63 s.

After around 65 s the main body wave phases have propagated out of the basin; however, large ground velocities still persist in the basin. In particular, the continuation of strong ground shaking is observed above the deepest parts of the basin (especially to the west and south-west of the label A and around label C in Figure 6.4, see also Figure 2.4).

To further examine the extension of ground motion for these parts of the basin we exemplarily apply the S transform, according to the description outlined in chapter 4.2.3.1, to the synthetic waveforms. Results for the installation site of station V1 are shown in Figure 6.5.

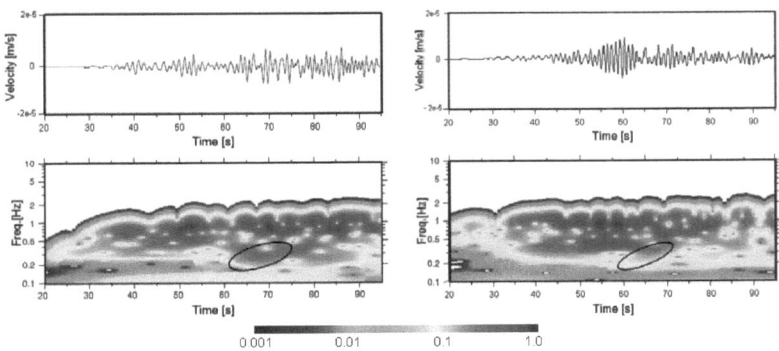

Figure 6.5: The vertical (left) and radial (right) component of the synthetic waveform (band-pass filtered between 0.05 and 1.8 Hz) for the 1 April 2010 earthquake at station V1 (top) and the corresponding normalized S transforms (bottom). The ellipse indicates possible dispersive waves.

The P-wave arrival at around 30 s is imaged correctly. Also the S-wave arrival at around 50 s can be identified by higher energy content in the seismogram, shown by dark red colours in the time frequency representation. Moreover, an arrival of a low-frequency wave train at around 60 s can be observed, especially on the vertical component. Similar to Figure 4.5, these waves follow the S-wave with a delay of around 10 s; in the seismogram, however, higher frequencies mask the exact arrival of this wave train. As described in chapter 4.2.3.2, these waves could be identified as secondary phases which might be generated at the basin edge by a conversion of S-waves. On the vertical component, the dispersive character of these waves is not as obvious as in Figure 4.5. On the other hand, on the radial, i.e. basin edge normal component, on which the Rayleigh wave should also be found, the ellipse in Figure 6.5 points out the existence of a dispersive wave train, but with significant lower energy.

Although this might not serve as a definitive proof for the existence of basin edge induced surface waves in the synthetic waveforms, there is anyhow a strong hint to the proper implementation of the sedimentary basin properties. Since we did not include further regional crustal heterogenities in our model, the existence of the observed low-frequency phases should depend only on the shape of the sediment bedrock interface and the impedance contrast, according to Narayan (2010). This further confirms the capability of the model to correctly simulate earthquake ground motion in the Santiago basin.

6.4. Simulating near-fault earthquake ground motion

6.4.1. Implementation

Since the accuracy of the implemented model has been validated and the influence of the low-speed sedimentary basin on wave propagation has been shown for events occurring well outside the basin, we now aim at quantifying the effect of near-fault earthquake ground motion on different assumptions, such as the role of hypocenter location. In this parametric analysis, we consider different locations of possible hypocenters along the San Ramón Fault (see chapter 2.2) which has been shown to pose a significant seismic hazard for the city. To this regard, a detailed mesh has been constructed which is depicted in Figure 6.6. It consists of 1,483,560 elements with a minimum size of about 30 m (inside the sedimentary basin) up to 900 m. The mesh will allow a simulation of frequencies up to around 2.5 Hz with a spectral degree of 4. The dynamic properties of the sedimentary basin and the bedrock layers are identical to the ones described in chapter 6.3.1.2.

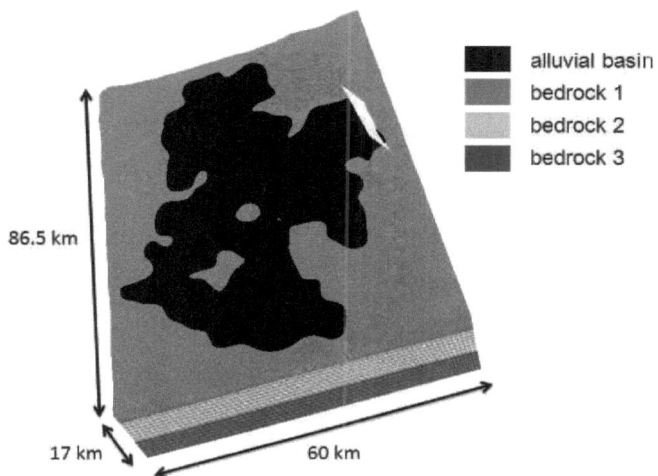

Figure 6.6: 3D hexahedral spectral element mesh adopted for scenario modeling. Different colors refer to different mechanical properties (see chapter 6.3.1.2). The shape of the Santiago basin is shown in black. The location of the San Ramón Fault is represented by a white rectangle.

All numerical simulations presented here correspond to earthquakes with a magnitude Mw=6.0 along the central segment of the San Ramón Fault, geometrically defined by a 12 km x 5 km rectangle (Wells and Coppersmith 1994). The parameters of the hypothetic source mechanism (strike=172°, dip=119°) are taken from Armijo et al. (2010) and are the same for all simulated scenario events. Like before, the rupture propagates circularly from the hypocenter with a constant rupture velocity of 2800 m/s. The slip velocity is around 1 m/s. For the G-ε and D-ε curves for the non-linear visco-elastic model adopted in this work, standard parameters for shallow soil materials were taken from Seed et al. (1986).

6.4.2. Influence of hypocenter location

The location of the hypocenter is regarded as one of the key factors affecting the distribution of ground motion and peak ground velocity (PGV) in the near-fault zone of an earthquake (Sommerville et al. 1997, Guatteri et al. 2003). Changing the hypocenter position will affect the relative time at which different parts of the fault radiate and how waves radiated during slip will interfere with each other. To study the effect of ground motion in the Santiago basin on this parameter, six different hypocenters were selected inside the same fault plane. For all the events, the seismic rupture is assumed to start from one hypocenter and then to propagate circularly along the fault. Figure 6.7 shows the source model corresponding to different hypocenter locations.

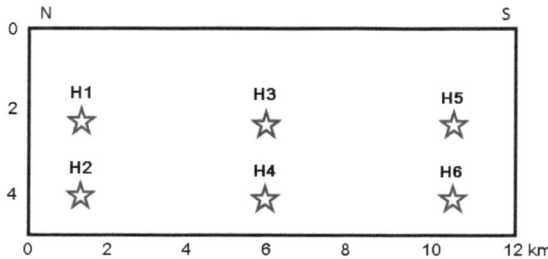

Figure 6.7: Hypocenter locations adopted for the parametric study.

As an example, Figure 6.8 presents snapshots of the velocimetric ground motion obtained by the H1 source model. Because the epicentre is located close to the basin, source radiation dominates the near field records. However, the interaction between the hypocenter location (and the corresponding directivity effects) and the surface topography in combination with the low wave velocities of the Santiago basin can be identified. After 6.0 s the rupture propagates radially from the starting point at the eastern basin edge. From 9.0 s to 12.0 s the propagation further into the basin continues. Compared to the portion of waves propagating in the western direction, the south-western and in particular the north-western wave packets have much larger amplitudes and delay due to the thick sediments (for the thickness of the sediments see Figure 2.4). At 13.5 s direct waves propagating toward north-west are reflected at the Cerro San Cristóbal (letter A in Figure 6.8), coinciding with later arrivals along a narrow zone to the south-east of the hill. A zone of strong ground motion parallel to the hill can be identified. In general, the width of such amplification zones and their distances from the basin edge depend on several parameters like the frequency content of the incident waves, the exact basin edge geometry, and the velocity contrast between the sediments and the bedrock (Pitarka et al. 1998). In the following snapshots, the prograding of the waves toward the north-west is clearly influenced by the Cerro Renca: low amplitudes

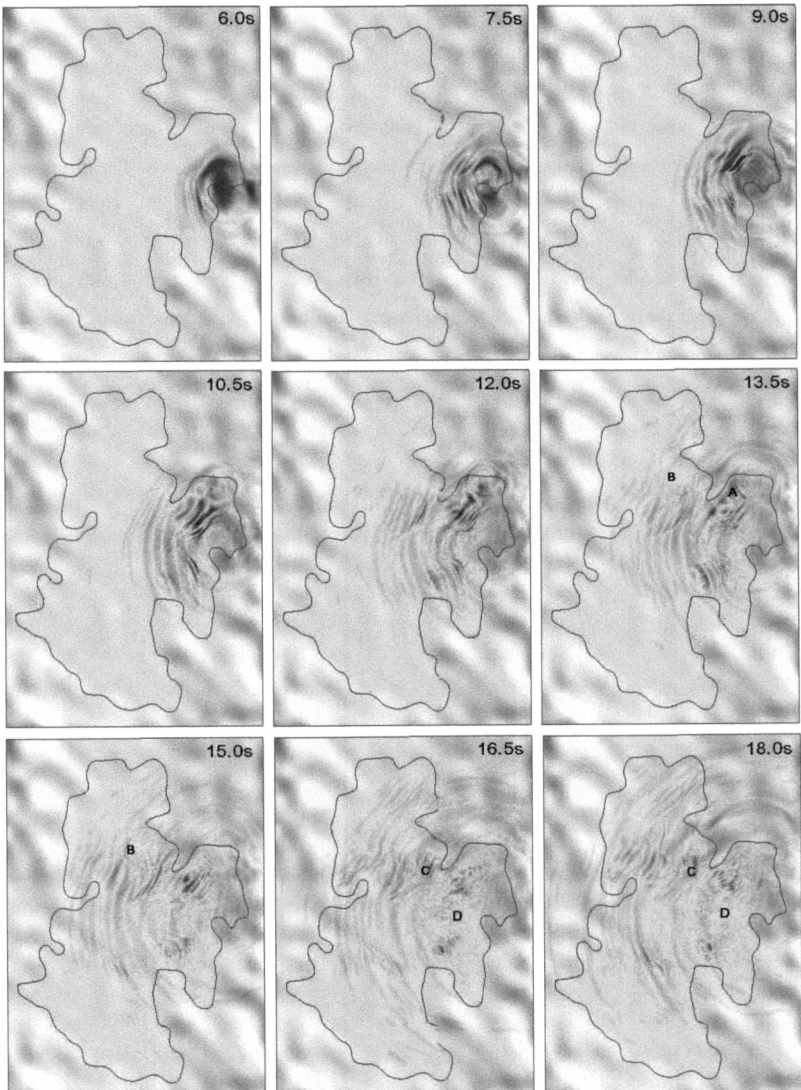

Figure 6.8: Snapshots of the absolute velocity of ground motion for the scenario earthquake with hypocenter H1. Red colors indicate positive velocities and blue colors indicate negative velocities. See text for further discussion.

are found where the hill is located (letter B), whereas high amplification can be seen southly. Interference effects and trapping of energy at distances of around 15 km from the fault (letter C) can be seen. The ground motion is as large as around 8 to 10 km west of the fault (letter D). Such trapping effects can also be seen in the waves travelling to the south-west but are less significant there.

Figure 6.9 shows the influence of hypocenter location on the maximum value of the PGV vector determined from the synthetic waveforms for each location. The amplitude of the velocity vector is the square root of the sum of the squares of the velocities for the three components. For deep structures like the Santiago basin, PGV rather than parameters of acceleration serves as the most relevant damaging potential indicator. In particular, the correlation of PGV with damage at low to intermediate frequencies has been found to be very good (see e.g. Brun et al. 2004 and references therein).

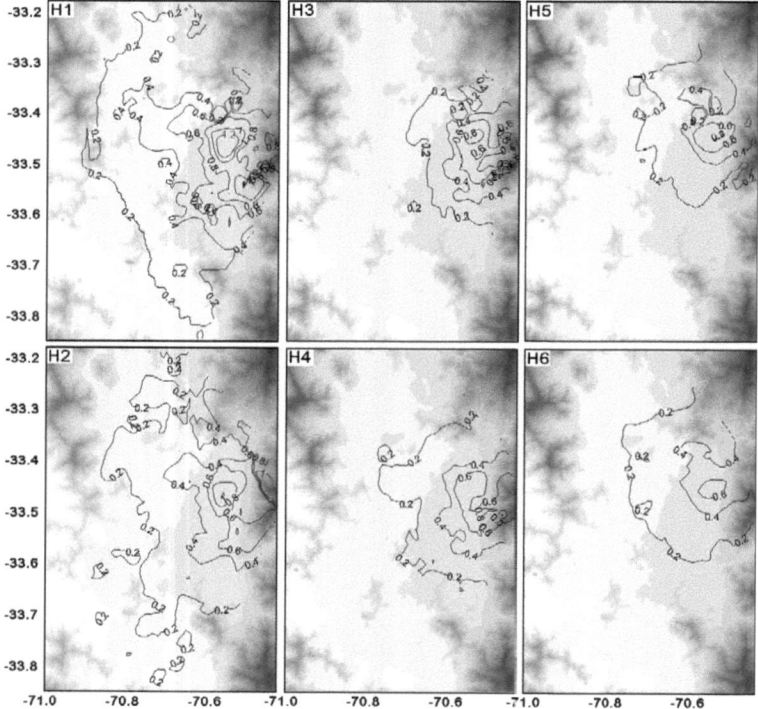

Figure 6.9: PGV for the various hypocenter locations shown in Figure 6.7. Isoline values are given in m/s.

In general, for cases in which the earthquake occurs on a fault at the basin margin, as has already been observed during the Hyogo-ken Nanbu (Kobe) earthquake, the influence of the basin becomes very important (e.g. Pitarka et al. 1998) causing particularly large-amplitude ground velocities. Corresponding to the various hypocenter locations and corresponding directivity effects, PGVs vary up to a factor of around 2. In general, the north-western part of the basin suffers most from the coupling of directivity, basin-edge effects, and interference with the highest PGV values found for all hypocenter locations on the top of a more than 400 m thick sedimentary layer. The worst case is hypocenter H1, located relatively shallow at the northern end of the fault, showing PGV values of up to 1.26 m/s. Also in the north-western part of the basin, where the sedimentary cover is much thicker, considerably higher PGV values (> 0.4 m/s) are found compared to the south-western part. On the contrary, for hypocenter H6 on the southern end of the fault maximum PGV values are only 0.65 m/s.

6.4.3. Discussion

The results of the simulations point out the influence of the low wave speed sedimentary basin. It is furthermore obvious that the pronounced topography can produce complex wave propagation behavior, with seismic energy reflected and scattered by the mountains. The combination of these effects as well as the radiation mechanism and hypocenter location with corresponding directivity effects play a relevant role in near-field ground motion. In particular, the unfavorable interaction between these parameters may raise concern about large PGV values even from low to moderate seismic events.

Anyhow, as can be seen in Figure 6.9, the high variability in PGV very close to the fault as a result from the simulations implies that exact prediction of ground motion in the near source region is subject to a high degree of uncertainty and is highly event specific. In reality, the specific complexities in the spatial and the temporal variation of the slip distribution on the fault will in all likelihood be different than assumed here. Even small variations in the rupture models may substantially affect the absolute PGV values. To this regard, it is also interesting to recall here that McGarr and Fletcher (2007), based on Brune (1970) and Ida (1973), found that the peak slip rate within the near-fault zone of an earthquake does not show any clear scaling with seismic moment or magnitude. They even concluded the existence of an upper bound of the PGV values independent of earthquake's magnitude and that this bound is controlled primarily by the source mechanism and the strength of the seismogenic crust. However, the values presented here do for sure not form an upper bound.

For sake of simplicity we further assumed a constant rupture velocity. In contrast, irregularities in the rupture velocity have been found to be the predominant source of high-frequency radiation (e.g. Madariaga 1977, Spudich and Frazer 1984). Yet, we assume that this effect might be of no consequence because of the 2.5 Hz maximum frequency limit considered for the simulations. Moreover, this limit might also be too low to actually account for strong lateral velocity variations in the sediments and shallow weathered layers in the bedrock.

Although the absolute values shown here should not be taken for granted the conclusion of this study is that directivity effects of the possible rupture of the San Ramón Fault might produce large ground motions in a very dense populated area in the north-eastern part of the Santiago basin. This part of the city is particularly endangered as the business district and a large part of the city's offices can be found there.

Although considering non-linear behaviour of the soil through a non-linear visco-elastic model, its role for the scenario simulations has been found to be less important than source mechanism and hypocenter location (e.g. Ambraseys et al. 2005, Stupazzini et al. 2009). This effect generally shows itself only in terms of a change in duration of ground motion in the strong motion part of the signal. Therefore, major variations of PGV in the western part of the basin far away from the source found in this study are more likely to be related to the late arrival of basin edge induced surface waves (Paolucci and Pitilakis 2007). This can also be seen in Figure 6.8 in which, for the south-western part of the basin, highest amplitudes are generated by later phases. However, the amplitude of these waves is highly sensitive to the correct implementation of the impedance contrast and the configuration of the sediment bedrock interface.

A more robust characteristic of the rupture process, in combination with high frequency simulations, has the potential to improve the modeling of near-fault ground motion. In turn, it should then be possible to improve our ability to predict the dynamic response of structures close to the source.

Chapter 7

Conclusions

The investigation carried out in this work showed that site effects have a considerable influence on earthquake ground motion in the basin of Santiago de Chile. This observation is consistent with the variable damage distribution in the city observed following the 1985 Valparaiso and the 2010 Maule earthquakes.

For the first time, a network of seismic stations was installed in the city of Santiago for recording earthquake signals. Time domain analysis shows differences in amplitude and lengthening with respect to a nearby rock site up to a factor of more than two, respectively. In particular, it could be shown that secondary phases (dispersive surface waves) are present and significantly increase the duration of shaking at the basin sites.

In the frequency domain, three different techniques for estimating site effects at the locations of the seismic stations have been applied. The comparison of the NHV ratio with EHV and SSR results for the sites of the seismic network illustrates a generally good agreement in the shape of all three curves, especially for the peak of the fundamental resonance frequency. The results obtained using P-wave windows often share similar features with those for S-wave windows because the frequencies of maximum amplification are almost the same, but often have lower amplitudes. Significant amplification (i.e. amplitudes exceeding 2 for the fundamental frequency) was found to be not only depend on the thickness of the sedimentary cover, but highly depends on local geological conditions and mechanical characteristics of the underlying material that can vary rapidly on short scale. Higher harmonics, which may overlap with the resonance frequencies of neighboring buildings, are often not visible in the NHV spectra, but they may not necessarily be damped out under such conditions. Therefore, NHV spectra might only serve as the lower frequency bound for site amplifications. Nonetheless, at some stations the amplitudes are quite small. We could also show that the NHV spectral ratios are quite stable with almost no dependency on the time and environment of the recordings.

Additionally, 146 measurements of seismic noise were carried out and analysed according to the SESAME criteria (Bard and SESAME WP02 team 2005). If a clear NHV peak appears, a confident estimate of the fundamental resonance frequency is possible. However, it could also be shown that for a case like Santiago de Chile, NHV curves not fulfilling the strict criteria may also consistently provide the soil resonance frequency or at least the frequency bands prone to amplification, especially when additional information on the geology is available. These results suggest that the quality criteria might sometimes be too restrictive. A similar conclusion was also reached by Bonnefoy-Claudet et al. (2009). On the other hand, the results highlight the importance of combining noise measurements with recordings of seismic events at characteristic sites for constraining the results.

Interpolation between the peaks of the NHV spectral ratios enabled us to map the fundamental resonance frequency of the investigated area. We showed that peaks mainly occur at low frequencies below 1 to 2 Hz, but slight amplification also affects frequencies from 2 Hz to 13 Hz. A general trend in the variation in the frequency range of amplification and the thickness of the sedimentary cover can be seen. Moreover, a correlation between the damage distribution in the city following the 2010 Maule earthquake and the fundamental frequency at some specific sites due to resonance effects could be pointed out.

For 125 of 146 noise measurements a further analysis was possible; the calculated H/V spectra were systematically inverted under the assumption of a horizontally layered one-dimensional structure below the measurement sites using geotechnical data and knowledge of the thickness of the sedimentary cover which had been determined previously by gravimetric measurements. This enabled us to derive S-wave velocity profiles for each measurement site which formed the basis for a 3D S-wave velocity model for the entire investigated area with the help of a kriging algorithm. Comparison of the results with existing velocity profiles gained by seismic refraction experiments and additional geological data shows good agreement. We found quite different S-wave velocity-depth gradients within the Santiago basin. However, the resolution of our model could be improved by adding more measurement sites.

As inexpensive and simple techniques for deriving local site response and amplification are of broad interest, we tested if a simple relationship between the slope of the topography and v_s^{30} exists on a local scale. Since no simple correlation between the topographic gradient and surficial geology (i.e. fast and more competent material indicating steeper slopes) exists for the investigated area, no linkage between slope and v_s^{30} could be found due to rather high S-wave velocities appearing in the Santiago basin, regardless of which spatial resolution is used. Although the investigated area is characterised by low and steep slope values, it is not possible to establish any estimation for site amplification based on topographic gradient. On the contrary, a better, but still only very approximate, correlation between v_s^{30} and local geological conditions is visible. In fact, we even showed that some geology can have different velocities and site amplifications. A higher resolution of the v_s^{30} map may, of course, be accomplished by adding more measurement sites. When taking into account the distribution of the mapped intensities of the 1985 Valparaiso event it could be illustrated that the intensity values are clearly influenced by the v_s^{30} distribution and cannot, even at a rudimentary level, be derived from the topographic gradient.

In addition to near-surface site conditions, seismic waves in basins are also known to be strongly influenced by thickness of the sedimentary cover and the shape of the sediment bedrock interface. To this regard, we built up a large-scale numerical model of the Santiago basin. The simulation of a regional event, which had been registered by a dense network installed in the city of Santiago for recording aftershock activity following the 27 February 2010 Maule earthquake, shows that the model, although simplified, is in general capable for realistic calculations of ground motion in terms of amplitude, duration, and frequencies below 1.8 Hz. Moreover, the proper implementation of topography and basin shape turns out to play a major role to adequately reproduce the observed effects of amplification of ground motion.

We also examined near-fault earthquake ground motion for a hypothetical event occurring along the active San Ramón Fault, which is crossing the eastern outskirts of the city. Among the possible factors contributing to ground motion variability we investigated the hypocenter location and corresponding directivity effects. Peak values of ground velocity may vary up to a factor of 2. Of course, other factors like slip velocity and rupture velocity will also significantly influence ground motion. Although the assumed source mechanism might too simple, we could show that the unfavorable interaction between fault rupture, radiation mechanism, and complex geological conditions in the near-field may give rise to large values of peak ground velocity even from moderate earthquakes. Therefore, owing to the available scanty near-fault strong motion data, 3D numerical simulations based on realistic underlying models presently serve as a valuable tool to provide estimations of near-fault ground motion.

Altogether, the combined use of experimental field studies and numerical calculations has been promising, suggesting that they can provide complementary information improving our ability to understand site and basin response. The results presented here might serve as a basis for the city's earthquake hazard assessment and may also further assist in risk reduction programs.

Appendix

Table A.1: Events recorded by seismic network between 21 March and 26 May 2008. Parameters are taken from PDE catalog. Shaded elements indicate that the station had not been installed at that time.

date	origin time (UTC)	latitude	longitude	depth [km]	magnitude	S1	S2	S3	S4	S5	S6	S7	S8
0321	111353	-33.04	-71.36	52	3.1	+	+						
0323	175731	-31.48	-70.45	118	4.0	+	+	+		+	+		
0324	203907	-20.04	-68.96	120	6.2	+	+		+	+	+		
0325	044022	-32.71	-71.68	11	3.1	+	+	+		+	+		
0325	045312	-31.65	-69.66	140	3.7	+	+	+		+	+		
0326	084624	-32.44	-71.51	45	4.0	+	+	+		+	+		
0330	125659	-32.96	-71.06	66	3.2	+	+	+		+	+	+	+
0330	181358	-25.91	-69.91	76	4.7	+	+	+	+	+	+	+	+
0401	143817	-31.83	-72.09	10	4.1	+	+	+	+	+	+	+	+
0403	010021	-28.64	-71.46	32	5.1	+	+	+	+	+	+	+	
0403	072501	-27.53	-71.20	33	4.8	+	+	+		+	+	+	+
0403	172244	-32.49	-71.53	56	4.0	+	+	+	+	+	+	+	+
0407	121558	-30.89	-71.49	36	4.1	+	+	+		+	+	+	+
0407	173841	-27.69	-71.06	63	4.8	+	+	+		+	+	+	+
0408	082436	-25.73	-70.70	28	4.4	+	+	+	+		+	+	+
0412	003012	-55.66	-158.45	16	7.1	+	+	+		+	+	+	+
0412	212157	-32.52	-71.70	29	4.0	+	+	+	+	+	+	+	+
0414	094519	-56.02	-28.03	140	6.0	+	+	+		+	+	+	+
0415	030304	13.56	-90.60	33	6.1	+	+	+		+	+	+	+
0415	233906	-31.84	-70.54	42	3.9	+	+	+		+	+	+	+
0416	084701	-30.47	-71.63	68	4.4	+	+	+		+	+	+	+
0417	075718	-19.94	-70.80	27	5.1	+	+	+		+	+	+	+
0418	021715	-35.98	-72.37	35	4.7	+	+	+	+	+	+	+	+
0423	070627	-34.35	-70.82	103	4.6	+	+	+	+	+	+	+	+
0428	155755	-58.74	-24.71	35	6.1	+	+	+		+	+	+	+
0502	060137	-29.62	-70.94	43	4.4	+	+	+	+	+	+	+	+
0503	093738	-31.25	-68.76	155	3.7	+	+	+	+	+	+	+	+
0504	140732	-32.64	-71.60	22	3.6	+	+	+		+	+	+	+
0514	173450	-28.27	-67.47	125	4.6	+	+			+	+	+	
0515	142328	-57.91	-25.48	35	5.9	+	+	+		+	+	+	+
0516	020221	-31.08	-70.53	100	4.7	+	+	+		+	+	+	+
0516	234100	-32.12	-71.54	43	3.9	+	+	+		+			
0518	171449	-30.97	-71.82	35	5.0	+	+	+	+	+	+	+	+
0523	193534	7.31	-34.90	9	6.5	+	+	+			+	+	
0523	204616	-22.85	-68.82	100	5.2	+	+	+			+	+	
0524	024336	-41.96	-72.19	9	5.3	+	+	+	+	+	+	+	

0524	065322	-42.05	-72.01	10	4.9	+	+	+		+	+	+	
0525	132258	-31.48	-71.90	51	4.7	+	+	+	+	+	+	+	

References

Aagaard, B. T., J. F. Hall, T. H. Heaton (2001): Characterization of near-source ground motions with earthquake simulations, *Earthq. Spec.* **17**, 177-207

Adams, B., N. M. Osborne, J. J. Taber (2003): The basin-edge effect from weak ground motions across the fault-bounded edge of the Lower Hutt Valley, New Zealand, *Bull. Seism. Soc. Am.* **93**, 2703-2716

Aki, K. (1957): Space and time spectra of stationary stochastic waves, with special reference to microtremors, *Bull. Earthquake Res. Inst.* **35**, 415-456

Allen, T. I., D. J. Wald (2009): On the use of high-resolution topographic data as a proxy for seismic site conditions (v_s^{30}), *Bull. Seism. Soc. Am.* **99**, 935-943

Alvarado, P., S. Barrientos, M. Saez, M. Astroza, S. Beck (2009): Source study and tectonic implications of the historic 1958 Las Melosas crustal earthquake, Chile, compared to earthquake damage, *Phys. Earth Planet. Inter.* **175**, 26-36

Ambraseys, N.N., J. Douglas, P. Smit, S. K. Sarma (2005): Equations for the estimation of strong ground motions from shallow crustal earthquakes using data from Europe and the Middle East: Horizontal peak ground acceleration and spectral acceleration, *Bull. Earthq. Eng.* **3**, 1-53

Ampuero, A., M. Van Sint Jan (2004): Velocidades de onda medidas en Santiago con el ensayo de refracción sísmica, *5th Congreso Chileno de Ingeniería Geotécnica*, Universidad de Chile, Santiago, Chile

Anderson, J. G., Y. Lee, Y. Zeng, S. M. Day (1996): Control of strong motion by the upper 30 meters, *Bull. Seism. Soc. Am.* **86**, 1749-1759

Ansal, A. (2004): Recent advances in earthquake geotechnical engineering and microzonation, Kluwer Academic Publishers, Dordrecht, The Netherlands

Arai, H., K. Tokimatsu (2000): Effects of Rayleigh and Love waves on microtremor H/V spectra, *Proc. 12th World Conf. on Earthq. Eng.*, paper 2232

Arai, H., K. Tokimatsu (2004): S-wave velocity profiling by inversion of microtremor H/V spectrum, *Bull. Seism. Soc. Am.* **94**, 53-63

Araneda, M., F. Avendano, C. Merlo (2000): Gravity model of the basin in Santiago, Stage III, *9th Chilenian Geological Congress* **2**, Santiago, Chile, 404-408

Archuleta, R. J., P-C Liu, J. H. Steidl, L. F. Bonilla, D. Lavallée, F. Heuze (2003): Finite-fault site-specific acceleration time histories that include nonlinear soil response, *Phys. Earth Planet. Interiors* **137**, 153-181

Armijo, R., R. Rauld, R. Thiele, G. Vargas, J. Campos, R. Lacassin, E. Kausel (2010): The West Andean Thrust (WAT), the San Ramón Fault and the seismic hazard for Santiago (Chile), *Tectonics* **29**, TC2007, doi: 10.1029/2008TC002427

Askari, R., H. R. Siahkoohi (2007): Ground roll attenuation using the S and x-f-k transforms, *Geophys. Prospect.* **55**, 1-10

Astroza, M., J. Monge (1991): Seismic microzones in the city of Santiago. Relation damage-geological unit, *4th International Conference on Seismic Zonation*, Stanford, USA

Astroza M., M. Moroni, M. Kupfer (1993): Calificación sísmica de edificios de albañilería de ladrillo confinada con elementos de hormigón armado, *Proc. XXVI Jornadas Sudamericanas de Ingeniería Estructural* **1**, Montevideo, Uruguay

Baize, S., S. Rebolledo, J. Lagos, R. Rauld (2006): A first-order geological model of the Santiago basin, *Proc. 1906 Valparaíso Earthquake Centennial*, Valparaiso, Chile, paper NGT1-05

Bao, H. S., J. Bielak, O. Ghattas, L. Kallivokas, D. R. O'Hallaron, J. R. Shewchuk J. F. Xu (1989): Large-scale simulation of elastic wave propagation in heterogeneous media on parallel computers, *Comput. Meth. Appl. Mech. Eng.* **152**, 85-102

Bard, P. Y. (2004): The SESAME project (2004): An overview and main results, *Proc. 13th World Conference on Earthquake Engineering*, Vancouver, Canada, paper 2207

Bard, P. Y., M. Bouchon (1980): The seismic response of sediment-filled valleys. Part 1. The case of incident SH waves, *Bull. Seism. Soc. Am.* **70**, 1263-1286.

Bard, P. Y., M. Bouchon (1985): The two-dimensional resonance of sediment filled valleys, *Bull. Seism. Soc. Am.* **75**, 519-541

Bard, P. Y., J. P. Riepl-Thomas (2000): Wave propagation in complex geological structures and their effects on strong ground motion, in: E. Kausel, G. Manolis (eds.), Wave Motion in Earthquake Engineering, WIT Press, Southampton, Boston, 39-95

Bard, P. Y., SESAME WP02 team (2005): Guidelines for the implementation of the H/V spectral ratio technique on ambient vibrations – measurements, processing and interpretations, *European Comission – Research General Directorate Project No. EVG1-CT-2000-00026 SESAME*, **D23.12**

Barrientos, S., E. Vera, P. Alvarado, T. Monfret (2004): Crustal seismicity in central Chile, *Journal of South American Earth Sciences* **16**, 759-768

Beresnev, I. A., G. M. Atkinson (2002): Source parameters of earthquakes in eastern and western North America based on finite-fault modeling, *Bull. Seism. Soc. Am.* **92**, 695-710

Bielak, J., J. Xu, O. Ghattas (1999): Earthquake ground motion and structural response in alluvial valleys, *J. Geotech. Geoenvirom. Eng.* **125**, 413-423

Bielak, J., O. Ghattas, E. J. Kim (2005): Parallel octree-based finite element method for large-scale earthquake ground motion simulation, *Comput. Model. Eng. Sci.* **10**, 99-112

Bindi, D., S. Parolai, D. Spallarossa, M. Catteneo (2000): Site effects by H/V ratio: Comparison of two different procedures, *J. Earthq. Eng.* **4**, 97-113

Bindi D, S. Parolai, F. Cara, G. Di Giulio, G. Ferretti, L. Luzi, G. Monachesi, F. Pacor, A. Rovelli (2009): Site amplifications observed in the Gubbio Basin, central Italy: hints for lateral propagation effects, *Bull. Seism. Soc. Am.* **99**, 741-760

Bommer, J. J., A. Martínez-Pereira (1999): The effective duration of earthquake strong motion, *J. Earthq. Eng.* **3**, 127-172

Bonilla, L. F., J. H. Steidl, G. T. Lindley, A. G. Tumarkin, R. J. Archuleta (1997): Site amplification in the San Fernando valley, California: variability of site effect estimation using S-wave, coda, and H/V methods, *Bull. Seism. Soc. Am.* **87**, 710-730

Bonamassa, O., J. E. Vidale (1991): Directional site resonances observed from aftershocks of the 18 October 1989 Loma Prieta earthquake, *Bull. Seism. Soc. Am.* **81**, 1945-1957

Bonnefoy-Claudet, S., S. Baize, L. F. Bonilla, C. Berge-Thierry, C. R. Pasten, J. Campos, P. Volant, R. Verdugo (2009): Site effect evaluation in the basin of Santiago de Chile using ambient noise measurements, *Geophys. J. Int.* **176**, 925-937

Borcherdt, R. D. (1970): Effects of local geology on ground motion near San Francisco Bay, *Bull. Seism. Soc. Am.* **60**, 29-61

Borcherdt, R. D. (1994): Estimates of site-dependent response spectra for design (methodology and justification), *Earthq. Spec.* **10**, 617-654

Bouchon, M. (1973): Effects of topography on surface motion, *Bull. Seism. Soc. Am.* **63**, 615-622

Bravo, R. D. (1992): Estudio geofisico de los suelos de fundación para un zonificacion sísmica del area urbana de Santiago Norte, PhD thesis, Universidad de Chile, Santiago, Chile

Bray, J., D. Frost (2010): Geo-engineering reconnaissance of the February 27, 2010, Maule, Chile, earthquake, published online under http://www.geerassociation.org/GEER_Post%20EQ%20Reports/Maule_Chile_2010/Cover_Chile_2010.html, last accessed 29 October 2010

Brun M., J. M. Reyouard, L. Jezequel, N. Ile (2004): Damaging potential of low magnitude near-field earthquakes on low-rise shear walls, *Soil Dyn. Earthq. Eng.* **24**, 587-603

Brune, J. N. (1970): Tectonic stress and the spectra of seismic shear waves from earthquakes, *J. Geophys. Res.***75**, 4997-5009

Building Seismic Safety Council (BSSC) (2004): NEHRP recommended provisions for seismic regulations for new buildings and other structures, 2003 edition (FEMA 450), Building Seismic Safety Council, National Institute of Building Sciences, Washington, USA

Capdeville, Y., E. Chaljub, J. P. Vilotte, J. P. Montagner (2003): Coupling the spectral element method with a modal solution for elastic wave propagation in global Earth models, *Geophys. J. Int.* **152**, 34-67

Caserta, A., A. Rovelli, F. Marra, F. Bellucci (1998): Strong diffraction effects at the edge of the Colfiorito, central Italy, basin, *Second Int. Symp. on the Effects on Surface Geology on Seismic Motion* **2**, Balkema, Yokohama, Japan, 435-440

Castellaro, S, F. Mulargia (2009): Vs30 estimates using constrained H/V measurements, *Bull. Seism. Soc. Am.* **99**, 761-773

Castro, R. R., RESNOM working group (1998): P- and S-wave site response of the seismic network RESNOM determined from earthquakes of Northern Baja California, Mexico, *Pure Appl. Geophys.* **1052**, 125-138

Çelebi, M. (1987): Topographical and geological amplifications determined from strong-motion and aftershock records of the 3 March 1985 Chile earthquake, *Bull. Seism. Soc. Am.* **77**, 1147-1167

CEN (2003): Eurocode (EC) 8: Design of structures for earthquake resistance – Part 1 General rules, seismic actions and rules for buildings, *EN 1998-1*, Brussels, Belgium

Chaljub, E., Y. Capdeville, J. P. Vilotte (2003): Solving elastodynamics in a fluid-solid heterogeneous sphere: a parallel spectral element approximation on non-conforming grids, *J. Comput. Phys.* **187**, 457–491

Chaljub E., D. Komatitsch, J. P. Vilotte, Y. Capdeville, B. Valette, G. Festa (2007): Spectral Element Analysis in Seismology, in Advances in Wave Propagation in Heterogeneous Media, Ru-Shan Wu, Valérie Maupin (eds.), Advances in Geophysics **48**, Elsevier, Amsterdam, The Netherlands, 365-419

Chaljub, E., P. Moczo, S. Tsuno, P. Y. Bard, J. Kristek, M. Käser, M. Stupazzini, M. Kristekova (2010): Quantitative comparison of four numerical predictions of 3D ground motion in the Grenoble Valley, France, *Bull. Seism. Soc. Am.* **100**, 1427-1455

Chatelain, J. L., B. Guillier, F. Cara, A. M. Duval, K. Atakan, P. Y. Bard, SESAME WP02 team (2008): Evaluation of the influence of experimental conditions on H/V results from ambient noise recordings, *Bull. Earthq. Eng.* **6**, 33-74

Chávez-García, F. J., W. R. Stephenson, M. Rodríguez (1999): Lateral propagation effects observed at Parkway, New Zealand: a case history to compare 1D vs 2D effects, *Bull. Seism. Soc. Am.* **89**, 718-732

Cornou, C., P. Y. Bard, M. Dietrich (2003): Contribution of dense array analysis to identification and quantification of basin-edge induced waves. Part II: application to Grenoble basin (French Alps), *Bull. Seism. Soc. Am.* **93**, 2624-2648

Cornou, C., P. Guéguen, P. Y. Bard, E. Haghshenas (2004): Ambient noise energy bursts obeservation and modeling: Trapping of harmonic structure-soil induced-waves in a topmost sedimentary layer, *J. Seismol.* **8**, 507-524

Cotte, N., H. A. Pedersen, M. Campillo, V. Farra, Y. Cansi (2000): Of great-circle propagation of intermediate-period surface waves observed on a dense array in the French alps, *Geophys. J. Int.* **142**, 825-840

Cruz, E., R. Riddell, S. Midorikawa (1993): A study of site amplification effects on ground motions in Santiago, Chile, *Tectonophysics* **218**, 273-280

Del Gaudio, V., S. Coccia, J. Wasowski, M. R. Gallipoli, M. Mucciarelli (2008): Detection of directivity in seismic response from microtremor spectral analyis, *Nat. Hazards Earth Syst. Sci.* **8**, 751-762

Faccioli, E., F. Maggio, A. Quarteroni, A. Tagliani (1996): Spectral-domain decomposition methods for the solution of acoustic and elastic wave equation, *Geophys.* **61**, 1160-1174

Faccioli, E., F. Maggio, R. Paolucci, A. Quarteroni (1997): 2D and 3D elastic wave propagation by a pseudo-spectral domain decomposition method, *J. Seismol.* **1**, 237-251

Fäh, D., E. Rüttener, T. Noack, P. Kurspan (1997): Microzonation of the city of Basel, *J. Seism.* **1**, 87-102

Fäh, D., F. Kind, D. Giardini (2001): A theoretical investigation of average H/V ratios, *Geophys. J. Int.* **145**, 535-549

Fäh, D., F. Kind, D. Giardini (2003): Inversion of local S-wave velocity structures from average H/V ratios, and their use for the estimation of site-effects, *J. Seism.* **7**, 449-467

Fäh, D., S. Steimen, I. Oprsal, J. Ripperger, J. Wössner, R. Schatzmann, P. Kästli, I. Spottke, P. Huggenberger (2006): The earthquake of 250 A. D. In Augusta Raurica, A real event with 3D site effect?, *J. Seismol.* **10**, 459-477

Field, E. H. (1996): Spectral amplification in a sediment-filled valley exhibiting clear basin-edge induced waves, *Bull. Seism. Soc. Am.* **86**, 991-1005

Field, E. H., K. H. Jacob (1995): A comparison and test of various site-response estimation techniques, including three that are not reference-site dependent, *Bull. Seism. Soc. Am.* **85**, 1127-1143

Frankel, A. D. (1994): Implications of felt area-magnitude relations for earthquake scaling and the average frequency of perceptible ground motion, *Bull. Seism. Soc. Am.* **84**, 462-465

Frankel, A. D., D. Carver, E. Cranswick, T. Bice, R. Sell, S. Hanson (2001): Observations of basin ground motions from a dense seismic array in San Jose, California, *Bull. Seism. Soc. Am.* **91**, 1-12

Frankel, A. D., D. L. Carver, R. A. Williams (2002): Nonlinear and linear site response and basin effects in Seattle for the M=6.8 Nisqually, Washington, earthquake, *Bull. Seism. Soc. Am.* **92**, 2090-2109

Frischknecht, C., P. Rosset, J. J. Wagner (2005): Towards seismic microzonation-2D modeling and ambient seismic noise measurements: the case of an embanked deep Alpine valley, *Earthq. Spec.* **21**, 633-651

Gaffet, S., C. Larroque, A. Deschamps, F. Tressols (1998): A dense array experiment for the observation of waveform perturbations, *Soil Dyn. Earthq. Eng.* **17**, 475-484

Gallipoli, M. R., M. Mucciarelli, M. Eeri, S. Gallicchio, M. Tropeano, C. Lizza (2004a): Horizontal to vertical spectral ratio (HVSR) measurements in the area damaged by the 2002 Molise, Italy, earthquake, *Earthq. Spec.* **20**, 81-93

Gallipoli, M. R., M. Mucciarelli, R. R. Castro, G. Monachesi, P. Contri (2004b): Structure, soil-structure response and effects of damage based on observations of horizontal-to-vertical spectral ratios of microtremors, *Soil Dyn. Earthq. Eng.* **24**, 487-495

Géli L., P. Y. Bard, B. Jullien (1988): The effect of topography on earthquake ground motion: a review and new results, *Bull. Seism. Soc. Am.* **78**, 42-63

GeoE-Tech (2007): Estudios sismicos en cuenca de Santiago perfiles de refracción sismica, Informe No. **2**, Santiago, Chile

Godoy, E., G. Yañez, E. Vera (1999): Inversion of an Oligocene volcano-tectonic basin and uplift of its superimposed Miocene magmatic arc, Chilean central Andes: First seismic and gravity evidence, *Tectonophysics* **306**, 217-326

Goodyear, B. G., H. Zhu, R. A. Brown, J. R. Mitchell (2004): Removal of phase artifacts from fMRI data using a Stockwell transform filter improves brain activity detection, *Magn. Reson. Med.* **51**, 16-21

Goovaerts, P. (2000): Geostatical approaches for incorporating elevation into the spatial interpolation of rainfall, *J. Hydrol.* **228**, 113-129

Gosar, A. (2008): Site effects study in a shallow glaciofluvial basin using H/V spectral ratios from ambient noise and earthquake data: The case of Bovec basin (NW Slovenia), *J. Earthq. Eng.* **12**, 17-35

Graves, R. W. (1993): Modeling threedimensional site response effects in the Marina District Basin, San Francisco, California, *Bull. Seism. Soc. Am.* **83**, 1042-1063

Graves, R. W. (1996): Simulating seismic-wave propagation in 3-D elastic media using staggered-grid finite-differences, *Bull. Seismol. Soc. Am.* **86**, 1091-1106

Graves, R. W., D. J. Wald (2004): Observed and simulated ground motions in the San Bernardino basin region for the Hector Mine, California, earthquake, *Bull. Seism. Soc. Am.* **94**, 131-146

Guatteri, M., P. M. Mai, G. C. Beroza, J. Boatwright (2003): Strong ground motion prediction from stochastic-dynamic source models, *Bull. Seism. Soc. Am.* **93**, 301-313

Guendelman, T. (2000): Perfil bio-sísmico de edificios. Un instrumento de calificación sísmica, *Revista BIT* **7**, Santiago, Chile, 30-33

Haghshenas, E., P. Y. Bard, N. Theodulis, SESAME WP04 team (2008): Empirical evaluation of microtremor H/V spectral ratio, *Bull. Earthq. Eng.* **6**, 75-108

Hall, J., T. Heaton, M. Halling, D. Wald (1995): Near-source ground motion and its effects on flexible buildings, *Earthq. Spec.* **11**, 569-605

Harkrider, D. G. (1964): Surface waves in multilayered elastic media, part 1, *Bull. Seism. Soc. Am.* **54**, 627-679

Harmsen, S., S. Hartzell, P. Liu (2008): Simulated ground motion in Santa Clara Valley, California, and vicinity from M≥6.7 scenario earthquakes, *Bull. Seism. Soc. Am.* **98**, 1243-1271

Hartzell, S., A. Leeds, A. Frankel, R. Williams, J. Odum, W. Stephenson, W. Silva (2002): Simulation of broadband ground motion including nonlinear soil effects for a magnitude 6.5 earthquake on the Seattle fault, Seattle, Washington, *Bull. Seism. Soc. Am.* **92**, 831-853

Hartzell, S., S. Harmsen, R. A. Williams, D. Carver, A. Frankel, G. Choy, Pengcheng Liu, R. C. Jachens, T. M. Brocher, C. M. Wentworth (2006): Modeling and validation of a 3D velocity structure for the Santa Clara Valley, California, for seismic wave simulations, *Bull. Seism. Soc. Am.* **96**, 1851-1881

Hatayama, K., K. Matsunami, T. Iwata, K. Irikura (1995): Basin-induced Love waves in the eastern part of the Osaka Basin, *J. Phys. Earth* **43**, 131 -155

Hisada, Y., K. Aki, T. L. Teng (1993): 3-D simulations of surface wave propagation in the Kanto sedimentary basin, Japan. Part 2: Application of the surface wave BEM, *Bull. Seism. Soc. Am.* **83**, 1700-1720

Holzer, T. L., A. C. Padovani, M. J. Bennett, T. E. Noce, J. C. Tinsely (2005): Mapping v_s^{30} site classes, *Earthq. Spec.* **21**, 353-370

Horike, M., H. Uebayashi, Y. Takeuchi (1990): Seismic response in three-dimensional sedimentary basin due to plane S-wave incidence, *J. Phys. Earth* **38**, 261-284

Horike, M., B. Zhao, H. Kawase (2001): Comparison of site response characteristics inferred from microtremor and earthquake shear waves, *Bull. Seism. Soc. Am.* **91**, 1526-1536

Ibs-von Seht, M., J. Wohlenberg (1999): Microtremor measurements used to map thickness of soft sediments, *Bull. Seism. Soc. Am.* **89**, 250-259

Ida, Y. (1973): The maximum acceleration of strong ground motion, *Bull. Seism. Soc. Am.* **63**, 959-968

Idei, T., M. Horike, T. Iwata (1985): Seismic coda waves observed on a sedimentary basin, *J. Seism. Soc. Japan* **38**, 217-232

International Conference on Building Officials (1997): Uniform Building Code, International Conference on Building Officials, Whittier, California, USA

Iriarte, S. (2003): Impact of urban recharge on long-term management of Santiago Norte aquifer, Santiago, Chile, Master thesis, Waterloo University, Ontario, Canada

Iriarte, S., M. Atenas, E. Aguirre, C. Tore (2006): Aquifer recharge and contamination determination using environmental isotopes: Santiago basin, Chile: A study case, Estudios de hidrología isotópica en América latina, International Atomic Energy Agency, Section Isotope geology, Vienna, Austria, 97-112

Isaaks, E. H., R. M. Srivastava (1989): Applied geostatics, Oxford University Press, New York, USA

Kagawa, T., S. Sawada, Y. Iwasaki (1992): On the relationship between azimuth dependency of earthquake ground motion and deep basin structure beneath the Osaka Plain, *J. Phys. Earth* **40**, 73-83

Kamae, K., K. Irikura, A. Pitarka (1998): A technique for simulating strong ground motions from great earthquakes, *Bull. Seism. Soc. Am.* **88**, 357-367

Kawase, H., K. Aki (1989): A study on the response of a soft basin for incident S, P, and Rayleigh waves with special reference to the long duration observed in Mexico City, *Bull. Seism. Soc. Am.* **79**, 1361-1382

Kawase, H., T. Sato (1992): Simulation analysis of strong motions in Ashigara Valley considering one- and two-dimensional geological structures, *J. Phys. Earth* **40**, 27-56

Khazaradze G., J. Klotz (2003): Short and long-term effects of GPS measured crustal deformation rates along the South-Central Andes, *J. Geophys. Res.* **108**, 1-13

Kinoshita, S., H. Fujiwara, T. Mikoshiba, T. Hoshino (1992): Secondary Love waves observed by a strong motion array in the Tokyo lowlands, Japan, *J. Phys. Earth* **40**, 99-116

Kitsunezaki, C., N. Goto, Y. Kobayashi, T. Ikawa, M. Horike, T. Saito, T. Kurota, K. Yamane, K. Okuzumi (1990): Estimation of P- and S-wave velocities in deep soil deposits for evaluating ground vibrations in earthquake, *J. JSNDS* **9**, 1-17

Knopoff, L., T. Levshina, V. I. Keilis-Borok, C. Mattoni (1996): Increased long-range intermediate-magnitude earthquake activity prior to strong earthquakes in California, *J. Geophys. Res.* **101**, 5779-5796

Komatitsch, D., J. Tromp (1999): Introduction to the spectral-element method for 3-D seismic wave propagation, *Geophys. J. Int.* **139**, 806-822

Komatitsch, D., J. Tromp (2002a): Spectral-element simulations of global seismic wave propagation, Part I: Validation, *Geophys. J. Int.* **149**, 390-412

Komatitsch, D., J. Tromp (2002b): Spectral-element simulations of global seismic wave propagation, Part II: 3-D models, oceans, rotation, and gravity, *Geophys. J. Int.* **150**, 303-318

Komatitsch, D., J. P. Vilotte, R. Vai, J. M. Castillo-Covarrubias, F. J. Sanchez-Sesma (1999): The spectral element method for elastic wave equations. Application to 2D and 3D seismic problems, *Int. J. Numer. Meth. Eng.* **45,** 1139-1164

Komatitsch, D., Q. Liu, J. Tromp, P. Süss, C. Stidham, J. H. Shaw (2004): Simulations of ground motion in the Los Angeles basin based upon the spectral-element method, *Bull. Seism. Soc. Am.* **94**, 187-206

Konno, K., T. Ohmachi (1998): Ground-motion characteristics estimated from spectral ratio between horizontal and vertical components, *Bull. Seism. Soc. Am.* **88**, 228-241

Kosloff, R., D. Kosloff (1986): Absorbing boundaries for wave propagation problems, *Journal of Computational Physics* **63**, 363-376

Kramer, S. L. (1996): Geotechnical Earthquake Engineering, Prentice-Hall International Series in Civil Engineering and Engineering Mechanics, Prentice-Hall, Upper Saddle River, New Jersey, USA

Krige, D. G. (1951): A statistical approach to some mine valuations and allied problems at the Witwatersrand, Master thesis, University of Witwatersrand, Johannesburg, South Africa

Lagos, J. (2003): Ignimbrita Pudahuel: caracterización geológica geotécnica orientada a su respuesta sísmica, Master thesis, Universidad de Chile, Santiago, Chile

Langston, C. A. (1979): Structure under Mount Rainier, Washington – Inferred from teleseismic P-waves, *Geophys. Res. Lett.* **84**, 4749-4762

Laslett, G. M. (1994): Kriging and splines: An empirical comparison of their predicative performance in some applications, *Journal of the American Statistical Association* **89**, 391-409

Lebrun, B., D. Hatzfeld, P. Y. Bard (2001): A site effect study in urban area: experimental results in Grenoble (France), *Pure Appl. Geophys.* **158**, 2543-2557

Lee, S. J., H. W. Chen, Q. Liu, D. Komatitsch, B. S. Huang, J. Tromp (2008): Three-dimensional simulations of seismic-wave propagation in the Taipei basin with realistic topography based upon the spectral-element method, *Bull. Seism. Soc. Am.* **98**, 253-264

Lermo, J., F. J. Chavez-Garcia (1993): Site effect evaluation using spectral ratios with only one station, *Bull. Seism. Soc. Am.* **83**, 1574-1594

Lermo, J., F. J. Chavez-Garcia (1994): Are microtremors useful in site response evaluation?, *Bull. Seism. Soc. Am.* **84**, 1350-1364

Madariaga, R. (1977): High-frequency radiation from crack (stress drop) models of earthquake faulting, *Geophys. J. R. Astron. Soc.* **51**, 625-651

Mai, P. M. (2009): Ground-motion complexity and scaling in the near-field of earthquake ruptures, in: R. Meyers (ed.), Encyclopedia of Complexity and System Sciences, Springer, Berlin, Heidelberg, Germany, 4435-4474

Mai, P. M., P. Somerville, A. Pitarka, L. Dalguer, S. Song, G. Beroza, H. Miyake, K. Irikura (2006): On scaling of fracture energy and stress drop in dynamic rupture models: consequences for near-source ground-motions, *Earthquakes: Radiated Energy and the Physics of Faulting, Geophysical Monograph Series of the American Geophysical Union* **170**

Malagnini, L., P. Tricarico, A. Rovelli, R. B. Herrmann, S. Opice, G. Biella, R. de Franco (1996): Explosion, earthquake, and ambient noise recordings in a Pliocene sediment-filled valley: inferences on seismic response properties by reference- and non-reference-site techniques, *Bull. Seism. Soc. Am.* **86**, 670-682

McGarr A., J. B. Fletcher (2007): Near-fault peak ground velocity from earthquake and laboratory data, *Bull. Seism. Soc. Am.* **97**, 1502-1510

McNamara, D. E., R. P. Buland (2004): Ambient noise levels in the continental United States, *Bull. Seism. Soc. Am.* **94**, 1517–1527

Moczo, P., M. Lucká, J. Kristek, M. Kristeková (1999): 3D displacement finite differences and a combined memory optimization, *Bull. Seism. Soc. Am.* **89**, 69-79

Molnar, S., J. F. Cassidy (2006): A comparison of site response techniques using weak-motion earthquakes and microtremors, *Earthq. Spec.* **22**, 169-188

Moroni, M. O., M. Astroza, C. Acevedo (2004): Performance and seismic vulnerability of masonry housing types used in Chile, *Journal of Performance of Constructed Facilities*, ASCE **18**, No.3

Mpodozis, C., V. A. Ramos (1989): The Andes of Chile and Argentina, in: G. E. Ericksen, M. T. Cañas Pinochet, J. A. Reinemund (eds.), Geology of the Andes and its relation to hydrocarbon and mineral resources, *Earth Science Series* **11**, Circum-Pacific council for energy and mineral resources, Houston, USA

Mucciarelli, M., G. Monachesi (1998): A quick survey of local amplifications and their correlation with damage observed during the Umbrio-Marchesan (Italy) earthquake of September 26, 1997, *J. Earthq. Eng.* **2**, 325-337

Mucciarelli, M., M. R. Gallipoli, M. Arcieri (2003): The stability of the horizontal-to-vertical spectral ratio of triggered noise and earthquake recordings, *Bull. Seism. Soc. Am.* **93**, 1407-1412

Nakamura, Y. (1989): A method for dynamic characteristics estimation of subsurface using microtremor on the ground surface, *Quaterly Reports of the Railway Technical Research Institute* **30**, 25-33

Narayan, J. R. (2010): Effects of impedance contrast and soil thickness on basin-transduced Rayleigh waves and associated differential ground motion, *Pure Appl. Geophys.*, in press, doi: 10.1007/s00024-010-0131-z

Nogoshi, M., T. Igarashi (1970): On the propagation characteristics estimations of subsurface using microtremors on the ground surface, *J. Seismol. Soc. Japan* **23**, 264-280

Nogoshi, M., T. Igarashi (1971): On the amplitude characteristics of microtremor, *J. Seismol. Soc. Japan* **24**, 26-40

Ohrnberger, M., F. Scherbaum, F. Krüger, R. Pelzing, S. K. Reamer (2004): How good are shear wave velocity models obtained from inversion of ambient vibrations in the Lower Rhine Embayment (N.W. Germany)?, *Bollettino di Geofisica teorica ed applicata* **45**, 215-232

Olsen, K. B., R. Archuleta, J. Matarese (1995): Magnitude 7.75 earthquake on the San Andreas fault: Three-dimensional ground motion in Los Angeles, *Science* **270**, 1628-1632

Olsen, K. B., S. M. Day, J. B. Minster, Y. Cui, A. Chourasia, M. Faerman, R. Moore, P. Maechling, T. Jordan (2006): Strong shaking in Los Angeles expected from southern San Andreas earthquake, *Geophys. Res. Lett.* **33**, L07305, doi10.1029/2005GL025472

Ortigosa, P., R. Lástrico (1971): Efecto sísmico. En informe de mecánica de suelos, *Informe No. 1*, Metropolitano de Santiago, Sec. Mec. Suelos, IDIEM, Universidad de Chile, Santiago, Chile, 41-47

Panou A. A., N. Theodulidis, P. Hatzidimitriou, K. Stylianidi, C. B. Papzachos (2005): Ambient noise horizontal-to-vertical spectral ratio in site effects estimation and correlation with seismic damage distribution in urban environment: the case of the city of Thessaloniki (Northern Greece), *Soil Dyn. Earthq. Eng.* **25**, 261-274

Paolucci, R. (2002): Amplification of earthquake ground motion by steep topographic irregularities, *Earthq. Eng. Struc. Dyn.* **31**, 1831-1853

Paolucci, R., E. Faccioli, F. Maggio (1999): 3D response analysis of an instrumented hill at Matsuzaki, Japan, by spectral method, *J. Seism.* **3**, 191–209

Paolucci, R., K. Pitilakis (2007): Seismic risk assessment of underground structures under transient ground deformations, in: K. Pitilakis (ed.), Earthquake Geotechnical Engineering, Springer, Berlin, Heidelberg, Germany, 433-459

Park, S., S. Elrick (1998): Predictions of shear-wave velocities in southern California using surface geology, *Bull. Seism. Soc. Am.* **88**, 677-685

Parolai, S. (2009): Denoising of seismograms using the S transform, *Bull. Seism. Soc. Am.* **99**, 226-234

Parolai, S., D. Bindi, P. Augliera (2000): Application of the generalized inversion technique (GIT) to a microzonation study: numerical simulations and comparison with different site-estimation techniques, *Bull. Seism. Soc. Am.* **90**, 286-297

Parolai, S., P. Bormann, C. Milkereit (2001): Assessment of the natural frequency of sedimentary cover in the Cologne area (Germany) using noise measurements, *J. Earthq. Eng.* **5**, 541-564

Parolai, S., S. Richwalski, C. Milkereit, P. Bormann (2004a): Assessment of the stability of H/V spectral ratios from ambient noise and comparison with earthquake data in the Cologne area (Germany), *Tectonophysics* **390**, 57-73

Parolai, S., D. Bindi, M. Baumbach, H. Grosser, C. Milkereit, S. Karakisa, S. Zünbül (2004b): Comparison of different site response estimation techniques using aftershocks of the 1999 Izmit earthquake, *Bull. Seism. Soc. Am.* **94**, 1096-1108

Parolai, S., J. J. Galiana-Merino (2006): Effect of transient seismic noise on estimated of H/V spectral ratios, *Bull. Seism. Soc. Am.* **96**, 228-236

Parolai, S., S. Richwalski, C. Milkereit, D. Fäh (2006): S-wave velocity profiles for earthquake engineering purposes for the Cologne area (Germany), *Bull. Earthq. Eng.* **4**, 65-94

Pasten, C. R. (2007): Respuesta sismica de la cuenca de Santiago, PhD thesis, Universidad de Chile, Santiago, Chile

Pelties, C., M. Käser, V. Hermann, C. Castro (2010): Regular versus irregular meshing for complicated models and their effect on synthetic seismograms, *Geophys. J. Int.* **183**, 1031-1051

Phillips, W. S., S. Kinoshita, H. Fujiwara (1993): Basin-induced Love waves observed using the strong-motion array at Fuchu, Japan, *Bull. Seism. Soc. Am.* **83**, 64-84

Picozzi, M., S. Parolai, D. Albarello (2005): Statistical analysis of noise horizontal-to-vertical spectral ratios (HVSR), *Bull. Seism. Soc. Am.* **95**, 1779-1786

Pinnegar, C. R. (2006): Polarization analysis and polarization filtering of three-component signals with the time-frequency S transform, *Geophys. J. Int.* **165**, 596-606

Pinnegar, C. R., D. W. Eaton (2003): Application of the S-transform to pre-stack noise attenuation filtering, *J. Geophys. Res.* **108**, 2422

Pitarka, A., K. Irikura, T. Iwata, H. Sekiguchi (1998): Three-dimensional simulation of the near-fault ground motion for the 1995 Hyogoken Nanbu (Kobe), Japan, earthquake, *Bull. Seism. Soc. Am.* **88**, 428-440

Pitarka, A., K. Kamae, P. Somerville, Y. Fukushima, T. Uetake, K. Irikura (2000): Simulation of near-fault strong-ground motion using hybrid Green's function, *Bull. Seismol. Soc. Am.* **90**, 566-586

Priolo E., J. M. Carcione, G. Seriani (1994): Numerical simulation of interface waves by high-order spectral modeling techniques, *J. Acoust. Soc. Am.* **95**, 681-693

Rauld, R., R. Armijo, G. Vargas, R. Thiele (2008): Morphology, geometry and kinematics of the San Ramón fault crossing Santiago, Chile (33.5° S), *4th Alexander von Humboldt International Conference – The Andes: Challenge for geosciences*, Santiago, Chile, paper 212

Rebolledo, S., J. Lagos, R. Verdugo, M. Lara (2006): Geological and geotechnical characteristics of Pudahuel Ignimbrite, Santiago, Chile, *Proc. of the 10th Internacional Association for Engineering Geology and the Environment International Congress*, Nottingham, United Kingdom, paper 106

Ripperger, J., J. P. Ampuero, P. M. Mai, D. Giardini (2007): Earthquake source characteristics from dynamic rupture with constrained stochastic fault stress, *J. Geophys. Res.* **112**, B04311, doi: 10.1029/2006JB004515

Rovelli, A., L. Scognamiglio, F. Marra, A. Caserta (2001): Edge-diffracted 1-sec surface waves observed in a small-size intramountain basin (Colfiorito, central Italy), *Bull. Seism. Soc. Am.* **91**, 1851-1866

Sánchez-Sesma, F. J., F. Luzon (1995): Seismic response of three-dimensional alluvial valleys for incident P, S, and Rayleigh waves, *Bull. Seism. Soc. Am.* **85**, 269-284

Scherbaum, F., K. G. Hinzen, M. Ohrnberger (2003): Determination of shallow shear wave velocity profiles in the Cologne Germany area using ambient vibrations, *Geophys. J. Int.* **152**, 597-612

Schimmel, M., J. Gallart (2005): The inverse *S* transform with windows of arbitrary and varying shape, *IEEE Trans. Signal Process.* **53**, 4417-4422

Schimmel, M., J. Gallart (2007): Frequency-dependent phase coherence for noise suppression in seismic array data, *J. Geophys. Res.* **112**, B04303, doi: 10.1029/2006JB004680

Seed H. B., R. T. Wong, I. M. Idriss, K. Tokimatsu (1986): Moduli and damping factors for dynamic analyses of cohesionless soils, *J. Geotech. Eng. ASCE* **112**, 1016-1032

Semblat, J. F., A. M. Duval, P. Dangla (2002): Seismic site effects in a deep alluvial basin: numerical analysis by the boundary element method, *Comput. Geotech.* **29**, 573-585

Sepulveda, S. A., M. Astroza, E. Kausel, J. Campos, E. A. Casas, S. Rebolledo, R. Verdugo (2008): New findings on the 1958 Las Melosas earthquake sequence, central Chile: Implications for seismic hazard related to shallow crustal earthquakes in subduction zones, *J. Earthq. Eng.* **12**, 432-455

Somerville, P. G., N. F. Smith, R. W. Graves, N. A. Abrahamson (1997): Modification of empirical strong ground motion attenuation relations to include the amplitude and duration effects of rupture directivity. *Seismol. Res. Lett.* **68**, 199-222

Spudich, P., L. N. Frazer (1984): Use of ray theory to calculate high frequency radiation from earhquake sources having spatially variable rupture velocity and stress drop, *Bull. Seismol. Soc. Am.* **74**, 2061-2082

Stacey, R. (1988): Improved transparent boundary formulations for the elastic wave equation, *Bull. Seism. Soc. Am.* **78**, 2089-2097

Stern, C., H. Amini, R. Charrier, E. Godoy, F. Hervé, J. Varela (1984): Petrochemistry and age of rhyolitic pyroclastic flows which occur along the drainage valleys of the Río Cachapoal (Chile) and the Río Yaucha and Río Papagayos (Argentina), *Revista Geológica de Chile* **23**, 39-52

Stockwell, R. G., L. Mansinha, R. P. Lowe (1996): Localization of the complex spectrum: The S transform, *IEEE Trans. Signal Process.* **44**, 998-1001

Strollo, A., S. Parolai, K. H. Jäckel, S. Marzorati, D. Bindi (2008a): Suitability of short-period sensors for retreiving reliable H/V peaks for frequencies less than 1 Hz, *Bull. Seism. Soc. Am.* **98**, 671-681

Strollo, A., D. Bindi, S. Parolai, K. H. Jäckel (2008b): On the suitability of 1 s geophone for ambient noise measurements in the 0.1 – 20 Hz frequency range: experimental outcomes, *Bull. Earthq. Eng.* **6**, 141-147

Stupazzini, M (2004): A spectral element approach for 3D dynamic soil-structure interaction problems, PhD thesis, University of Milan, Italy

Stupazzini M., R. Paolucci, H. Igel (2009): Near-fault earthquake ground motion simulation in the Grenoble valley by a high-performance spectral element code, *Bull. Seism. Soc. Am.* **99**, 286-301

Teves-Costa, P., L. Matias, P. Y. Bard (1996): Seismic behaviour estimation of thin alluvial layers using microtremor recordings, *Soil Dyn. Earthq. Eng.* **15**, 201-209 (1996)

Theodulidis, N. P. (2006): Site characterization using strong motion and ambient noise data: Euroseistest (N Greece), *3rd Symposium on Effects of Surface Geology on Seismic Motion*, Grenoble, France, 801-810

Thiele, R. (1980): Geology of the Santiago metropolitan area, *Charter Geológica de Chile* **39**, Institute of Geological Research, Santiago, Chile, 51-52

Tokimatsu, K., Y. Miyadera (1992): Characteristics of Rayleigh waves in microtremors and their relation to underground structures, *J. Struct. Constr. Eng.* **439**, 81-87

Toshinawa, T., M. Matsuoka, Y. Yamazaki (1996): Ground-motion characteristics in Santiago, Chile, obtained by microtremor observations, *11th World Conference on Earthquake Engineering*, Paris, France, paper 1764

Trifunac, M. D., A. G. Brady (1975): A study on the duration of strong earthquake ground motion, *Bull. Seism. Soc. Am.* **65**, 581-626

Tromp, J., C. H. Tape, Q. Liu (2005): Seismic tomography, adjoint methods, time reversal, and banana-doughnut kernels, *Geophys. J. Int.* **160**, 195-216

Valenzuela, G. B. (1978): Suelo de fundación del Gran Santiago, Inst. Invest. Geol. Santiago, Chile, Bol. **33**

Vicente, J. C. (2005): Dynamic paleogeography of the Jurassic Andean Basin: Pattern of transgression and localisation of main straits through the magmatic arc, *Asoc. Geol. Argent. Rev.* **60**, 221-250

Vidale, V.E., D. V. Helmberger (1988): Elastic finite-difference modeling of the 1971 San Fernando California earthquake, *Bull. Seism. Soc. Am.* **78**, 122-141

Wackernagel, H. (1998): Multivariate statistics. An introduction with applications, Springer, New York, USA

Wald, D. J., R. W. Graves (1998): The Los Angeles basin response in simulated and recorded ground motions, *Bull. Seism. Soc. Am.* **88**, 337-355

Wald, D. J., T. I. Allen (2007): Topographic slope as a proxy for seismic site conditions and amplification, *Bull. Seism. Soc. Am.* **97**, 1379-1395

Wegner, J. L., M. M. Yao, X. Zhang (2005): Dynamic wave-soil-structure interaction analysis in the time domain, *Comp. and Struc.* **83**, 2206-2214

Wells, D., K. Coppersmith (1994): New empirical relationships among magnitude, rupture length, rupture width, rupture area, and surface displacement, *Bull. Seism. Soc. Am.* **84**, 974-1002

Wills, C. J., M. D. Petersen, W. A. Bryant, M. S. Reichle, G. J. Saucedo, S. S. Tan, G. C. Taylor, J. A. Treiman (2000): A site-conditions map for California based on geology and shear wave velocity, *Bull. Seism. Soc. Am.* **90**, 187-208

Wooleroy, E. W., R. Street (2002): 3D near-surface soil response from H/V ambient-noise ratios, *Soil Dyn. Earthq. Eng.* **22**, 865-876

Yalcinkaya, E., O. Alptekin (2005): Site effect and its relationship to the intensity and damage observed in the June 27, 1998 Adana-Ceyhan earthquake, *Pure Appl. Geophys.* **162**, 913-930

Yamanaka, H., H. Ishida (1996): Application of generic algorithms to an inversion of surface-wave dispersion data, *Bull. Seism. Soc. Am.* **86**, 436-444

Zhu, H., B. G. Goodyear, M. L. Lauzon, R. A. Brown, G. Mayer, A. G. Law, L. Mansinha, J. R. Mitchell (2003): A new local multiscale Fourier analysis for MRI, *Med. Phys.* **30**, 1134-1141

Zienckiewicz, O. C., R. L. Taylor (1989): The finite element method, McGraw-Hill, London, United Kingdom

Acknowledgements

I am indebted to many people for their long-lasting support and encouragement which was invaluable for the successful completion of this research work.

First of all I owe my deepest gratitude to Stefano Parolai for his guidance and patient support throughout these years. His ongoing encouragement is notably appreciated.

The second special thanks to my advisor Prof. Zschau for his kindness and his ability in finding the right words to any kind of scientific discussion.

Thanks to Marco Stupazzini in Munich and Roberto Parolucci and Chiara Smerzini at the Politecnico di Milano for their assistance in the numerical calculations and their warm hospitality.

Many words of gratitude to the people working here in Potsdam in the section for their help and encouragement, in particular to: Claus and Regina Milkereit, Susanne Köster, Erwin Günther and Birger Lühr.

I am indebted for many reasons to my friends and colleagues (Angelo, Dino, Domenico, Marc, Matteo, Max). I will always remember the very nice time spent together between Potsdam and Berlin.

In Chile, thanks to Natalia Silva, David Solans, Claudia Honores, José Gonzalez of the Universidad de Chile for supporting me in the field experiments.

I am grateful to Kevin Fleming who kindly improved my English.

Last but not least, I dedicate this thesis to the most important people in my life, my family for their unreserved love and support, and for being my source of inspiration and happiness during these years. There are not words to say thanks to them.

Die VDM Verlagsservicegesellschaft sucht für wissenschaftliche Verlage abgeschlossene und herausragende

Dissertationen, Habilitationen, Diplomarbeiten, Master Theses, Magisterarbeiten usw.

für die kostenlose Publikation als Fachbuch.

Sie verfügen über eine Arbeit, die hohen inhaltlichen und formalen Ansprüchen genügt, und haben Interesse an einer honorarvergüteten Publikation?

Dann senden Sie bitte erste Informationen über sich und Ihre Arbeit per Email an *info@vdm-vsg.de*.

Sie erhalten kurzfristig unser Feedback!

VDM Verlagsservicegesellschaft mbH
Dudweiler Landstr. 99
D - 66123 Saarbrücken

Telefon +49 681 3720 174
Fax   +49 681 3720 1749

www.vdm-vsg.de

Die VDM Verlagsservicegesellschaft mbH vertritt

Printed by Books on Demand GmbH, Norderstedt / Germany